7th Nanoscience and Nanotechnology Symposium (7th NNS 2017)

Better Life with Nanoscience and Nanotechnology

IOP Conference Series: Materials Science and Engineering
Volume 395

Bandung, Indonesia
22 - 24 October 2017

Editors:

Toto Winata
Ferry Iskandar
Satria Zulkarnaen Bisri
Akfiny H. Aimon

ISBN: 978-1-5108-7025-3
ISSN: 1757-8981

Printed from e-media with permission by:

Curran Associates, Inc.
57 Morehouse Lane
Red Hook, NY 12571

Some format issues inherent in the e-media version may also appear in this print version.

Printed by Curran Associates, Inc. (2018)

For permission requests, please contact the Institute of Physics
at the address below.

Institute of Physics
Dirac House, Temple Back
Bristol BS1 6BE UK

Phone: 44 1 17 929 7481
Fax: 44 1 17 920 0979

techtracking@iop.org

Additional copies of this publication are available from:

Curran Associates, Inc.
57 Morehouse Lane
Red Hook, NY 12571 USA
Phone: 845-758-0400
Fax: 845-758-2633
Email: curran@proceedings.com
Web: www.proceedings.com

Table of contents

Volume 395

7th Nanoscience and Nanotechnology Symposium (NNS)

22-24 October 2017, Bandung, Indonesia

Accepted papers received: 19 July 2018

Preface

Papers

IOP Conf. Series: Materials Science and Engineering **395** (2018) 011001 doi:10.1088/1757-899X/395/1/011001

PREFACE

The Nanoscience and Nanotechnology Symposium (NNS) conference series is one of The Materials Research Society of Indonesia (MRS-ID) international events that brings together researchers in the field of nanoscience and nanotechnology from Indonesia and abroad. These international biennial symposia have been held since 2008. In 2017, the Physics of Electronic Materials Division of Institut Teknologi Bandung was honored to serve as the organizer of the 7th Nanoscience and Nanotechnology Symposium (NNS).

The 7th Nanoscience and Nanotechnology Symposium (NNS) was held in Bandung from 22nd to 24th October 2018 with the central theme "Better Life with Nanoscience and Nanotechnology". It was co-organized by both The Materials Research Society of Indonesia (MRS-ID) and Faculty of Mathematics and Natural Sciences of Institut Teknologi Bandung (ITB).

This event aimed to support Indonesian researchers in the field of nanoscience and nanotechnology and enable them to develop supportive networks as well as to update themselves with the latest progress in the field of nanoscience and nanotechnology. We were pleased that this event was able to convene prominent scientists, researchers, and engineers from both Indonesia and abroad to share their latest scientific works and discoveries in the field of nanoscience and nanotechnology.

There were four plenary speakers, six invited speakers, 92 contributed speakers and numbers of listener participants. Based on the countries of their affiliations, they were from 5 (five) countries including Indonesia. The foreign participants came from Japan, Malaysia, Germany, and The Netherlands. This branded MRS-ID international event consisted of 7 (seven) symposia that cover the broad scope of nanoscience and nanotechnology.

Majority of the presenters in this meeting had submitted their full papers for possible publications in either the peer-reviewed conference proceeding (a proceeding curated by IOP Conference Series: Materials Science and Engineering) or in a peer-reviewed journal, Materials Research Express (MRX) (also published by IOP publishing). After selections and the peer-review process by experts in related fields, 26 manuscripts are accepted for publication in this proceeding, in addition to some numbers published in MRX.

The 7th Nanoscience and Nanotechnology Symposium (NNS) committee acknowledged the enthusiasms of the participants and numerous sponsors, all members of the program committee and organizing committee, the international advisory board, Faculty of Mathematics and Natural Sciences of ITB, executive committee of MRS-ID and all those who had contributed to the success of the 7th Nanoscience and Nanotechnology Symposium (NNS).

Prof. Toto Winata
Chairman of the 7th NNS 2017

Dr.rer.nat. Akfiny H. Aimon
Co-Chairman of the 7th NNS 2017

IOP Conf. Series: Materials Science and Engineering **395** (2018) 011002 doi:10.1088/1757-899X/395/1/011002

Peer review statement

All papers published in this volume of *IOP Conference Series: Materials Science and Engineering* have been peer reviewed through processes administered by the proceedings Editors. Reviews were conducted by expert referees to the professional and scientific standards expected of a proceedings journal published by IOP Publishing.

Effect of Microwave Irradiation on the Zeolite Properties Synthesized from Rice Husk Ash

P E Suryani and A D Candra

Ronggolawe College of Technology, Blora, Indonesia

Puputeka@sttrcepu.ac.id

Abstract. Indonesia is an agricultural country with the main product was rice. The abundance of rice farming contributes to the amount of rice husks every year. Today, the rice husk has been widely used as a source of silica in the preparation of synthetic zeolite in many researches. In this work, increasing characteristics of synthetic zeolite from rice husk ash by microwave irradiated method is investigated. The prepared zeolite samples were characterized using X-Ray Diffraction (XRD) to determine crystallinity and zeolite components. In application, synthetic zeolite is used in improvement of waste cooking oil quality. Waste cooking oil quality was determined by peroxide value. Modification of zeolite by microwave irradiation fixed in 100 W and 180 W variable of power for 60 minute. Both treated and untreated sample formed high crystallinity based on the XRD traces that is above 90%. In this project, the wasted cooking oil sample is taken from the rest of the frying fish. Adsorption was carried out at 60 °C with constant stirring in each sample. The peroxide value was analyzed by iodometric method, directly. The initial peroxide value of waste cooking oil is 9 meq/kg, this value decreased since adsorption process by prepared zeolite. Among the adsorption processes, 180 W modified zeolite yield highest decrease of peroxide value i.e 2 meq/kg, follow by 100 W modified which is 5 meq/kg, and the least decrease was reached by unmodified zeolite adsorption i.e 6 meq/kg.

1. Introduction

Paddy is a commodity with the largest amount of production in Indonesia compared to other food crops such as corn, soybeans, peanut, etc. The amount of paddy production tends to increase every year, significantly. This can be seen from the amount of rice production in 2015 that is reached 75 tons, while rice production in 2014 amounted to 70 tons [1]. The abundance of paddy in Indonesia will have a serious impact on the environmental pollution, due to rice husk from residue milling. Thus, potent technique should be applied to utilize rice husk, effectively. In the process of milling, paddies produce 20-30% of rice husk of total weight [2][3]. The result of burning rice husk as rice husk ash is a potential source of silica. Rice husk contain approximately 20-25% of silica [4][5]. Rice husk is one of the most underutilized biomass, most of the rice husk is just burned. Many studies have been conducted to increase the economic value of rice husk, so that rice husk can be used as one of renewable material sources. Hydrothermal is commonly used method used in the manufacture of synthetic zeolites. In the past year, a new technique of microwave-utilize has been widely introduced as a novel energy. By its properties, microwave is able to heat objects rapidly, quickly and effectively [6]. Several methods in the zeolite manufacturing process have been performed to produce zeolites with superior characteristics. Heat is regenerated thoroughly by rotate motion, it might cause the shorten heating time in chemical reactions [7][8]. Previous study mentioned that the synthesis process

IOP Conf. Series: Materials Science and Engineering **395** (2018) 012001 doi:10.1088/1757-899X/395/1/012001

of zeolite-A from CFA using microwave irradiation obtained shorten the crystallization time compared with conventional heating method i.e. from 8-89 hours to 10-30 minutes [9]. The microwave utilization also showed a positive response on the synthesis of zeolite Na-p1 from CFA [10]. Arafat et al [11] concluded that the process of synthesizing zeolite-Y and ZSM-5 by involving microwave result shorten time from 10-50 hours to 10 minutes on zeolite-Y and from several days to minutes (30 minutes) of zeolite ZSM-5. We have already known that amorphous silica hold great promise in promise in heavy metal stabilization [12]. In this work, the obtained both modified and unmodified zeolite were used to tread decreasing peroxide present in waste cooking oil. The effect of microwave irradiation on zeolite sorption capability will be investigate further.

2. Experimental Procedure

2.1 Sample Preparation

The initial stage in the synthesis of zeolites is initiated from the making of rice husk ash. Previously rice husk was thoroughly washed to remove soil and other impurities. Water contained in rice husks is evaporated by oven at 100°C for 4 h. The dried rice husk is then burned in the furnace for 3 hours at 600°C. The obtained rice husk ash is then washed with HCl to remove the metal content. After that, rice husk ash was washed by dissolving in distilled water until reach neutral pH. The neutral pH of rice husk ash let dried in an oven at 150°C for 4 h. The white ash product from this process was named silica.

2.2 Sodium Silicate Preparation

The neutral silica was pounded carefully to obtain fine particle [13]. After that, as much as 10 g of silica which was dissolved into 82.5 ml NaOH 4 M was boiled under constant stirring, and let the solution evaporated and form dark gel. Formed gel was poured to the petri dish thoroughly and then anneal it at elevated temperature (500°C) for 3 h. The final Filtrate that was produced in this step was called sodium silicate (Na_2SiO_3).

2.3 Sodium Aluminate Preparation

A number of sodium hydroxide (20 g) was incorporated to 100 ml distilled water, briefly. After that place the solution to the constant stirring for 1 h and 100°C heating. Then poured 8,5 g aluminium oxide to the sodium hydroxide solution gradually.

2.4 Zeolite Preparation

Both of Sodium aluminate and sodium silicate solution were blend homogeneously by constant stirring, homogeneous mixture was then fed into the microwave at various power of 100 W, 180 W, for 60 min. The unmodified sample was treated by stirred in constant stirring for 5 h, in 80°C of heating. The next stage was neutralized the zeolite using distilled water thoroughly. The neutral zeolite was then anneal at 400°C before it was characterized.

2.5 Characterization

X-ray diffraction is a common method for analyzing the structure of a crystal. In this study sample was analyzed by Shimadzu-7000. The testing was start by interacting the sample with electromagnetic waves of X-rays that have a wavelength λ= 0.5-2.5 Å [14].

3. Result and Discussion

Effect of microwave irradiation on zeolite morphology also has investigated in our previous study [15]. Compare the unmodified zeolite, treated zeolite by microwave irradiation proved to be more effective in formed crystalline structure in time preparation. Microwave irradiation process was carried out the object with rapidly and selectively heating process. It's one of equipment that can be rotate and spread out of the energy to the object. Table 1 shows the crystallinity of both modified and unmodified zeolite. Zeolite without irradiation had highest percentage of crystallinity compare others. In this case unmodified zeolite under 5 h stirring and heating, large more time compare to modified

IOP Conf. Series: Materials Science and Engineering **395** (2018) 012001 doi:10.1088/1757-899X/395/1/012001

zeolite that was needs 1 h preparation. Transformation silica contained in ash was deeply depend on annealing temperature. In this work sample of zeolite was burnt in temperature of 600-700°C, this condition was high enough to form crystalline structure on zeolite. This result slightly different which is explained by Nair and Co-workers, the amorphous silica can be produced by combustion process at 500-700°C, greater temperature was form crystallinity on zeolite [16]. This condition might be caused by different kind of paddy-utilize in the research.

Table 1. The crystallinity of modified and unmodified zeolite.

Sample of zeolite	**Crystallinity (%)**
Unmodified zeolite	93,684
Treated 100 W	90,807
Treated 180 W	90,212

Effect of irradiation power is studied in 0 W, 100 W, and 180 W, 1 h. Both sodium silicate and sodium aluminate were mixed in constant stirring for several minutes before heating process. The influence of microwave irradiation on the sample is analyzed by XRD method to determine the percentage of each component. High crystallinity was indicated by strong peak at 2θ: 10-11, 22-24 and 29-31. Auerbach et al [17] identified range of crystallinity zeolite with cube shape in measurement 2θ:7-8, 10-11, 22-24, and 29-31. The XRD date was analyzed by software Match to determine component of prepared zeolite. Untreating zeolite resulted in the least percentage of silica (3.3 %), Unmodified sample contain of 3.3% silica and 96.7% alumina, these indicate the blending process of sodium aluminate and sodium silicate produces less bind of each compound. It was slightly different to the treated zeolite that is resulting more contain of silica. Microwave irradiation of 100 W result of the highest (16.1 wt %) amount of SiO_2, follow by treated zeolite 180 W which is contain of 5.8 wt % of SiO2. The result of the investigation tells us that microwave irradiation contribute to the binding process among sodium aluminate and sodium silicate. This result shows microwave irradiation affect to the zeolite properties.

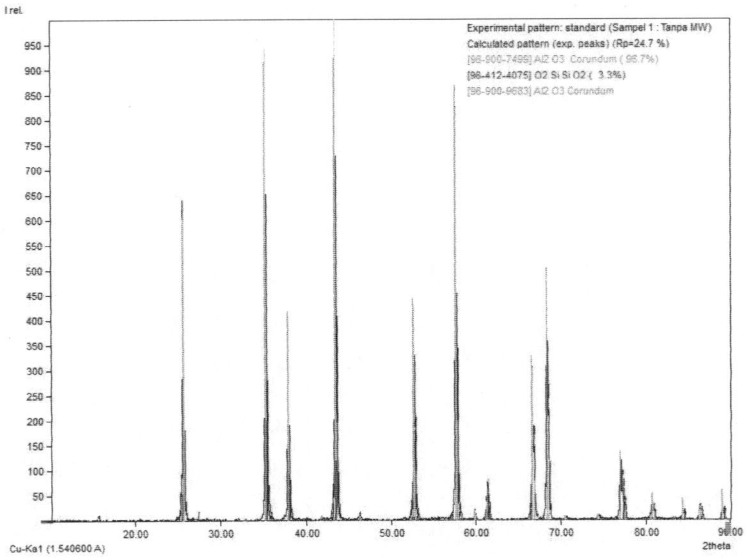

Figure 1. Unmodifed sample (0 W).

According to the previous studied, crystallinity forming was start at 60 min in zeolite preparation by microwave irradiation [18]. XRD trace of modified zeolite running under 100 W shown in Fig 2. Based on the strong peak of XRD diffractograms, sample resulting high crystallinity. According to the Match calculation, sample with 100 W irradiation for 1 h has highest silica value i.e. 16.1wt %. These conditions indicate that irradiation process gives some prospective method than the conventional method.

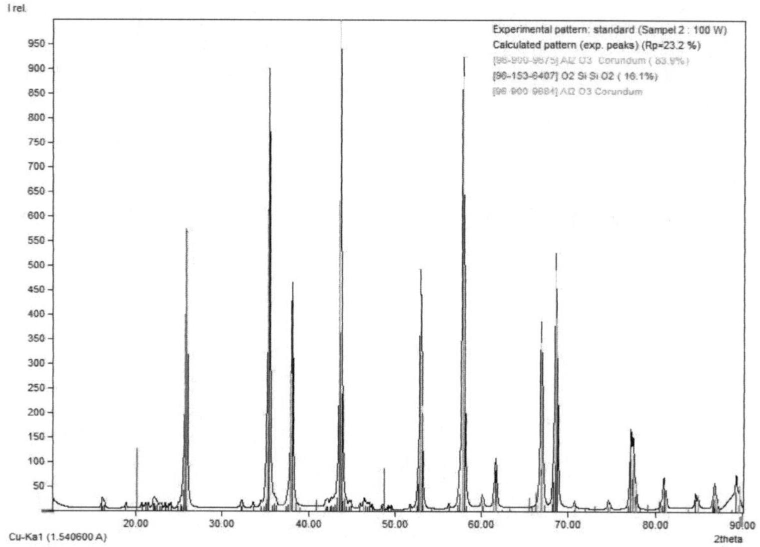

Figure 2. Modified Sample 100 W.

The main structure of modified sample 180 W can be seen in Fig 3. XRD diffractograms tells the same crystallinity of modified sample in 100 W. This result indicates that the power modification from 100 W to 180 W did not gave significant effect. The data from Match analyze give the fact that sample contains of 5.8% silica and 94.2% alumina. These might be caused by the leaching process to get a neutral pH.

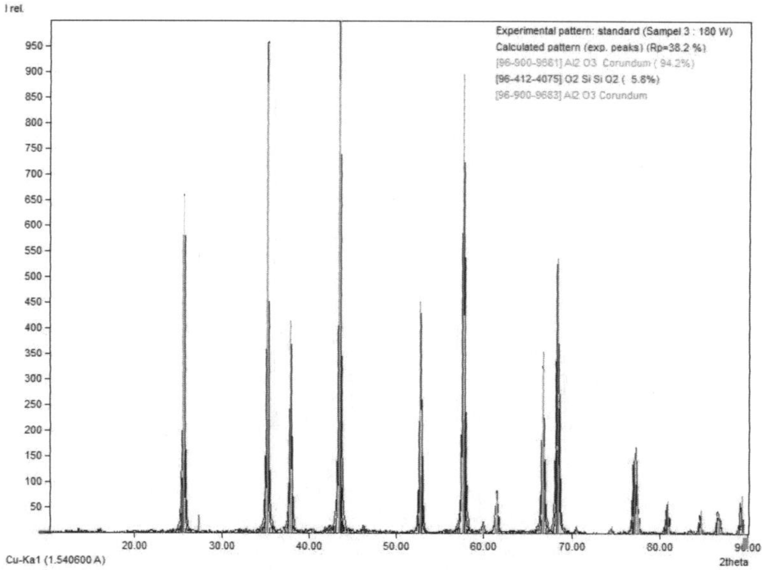

Figure 3. Modified sample 180 W.

XRD date showed that both modified and unmodified sample resulting high crystallinity. Furthermore, it is also found that these can be expressed by each single correlation line. Hence, the crystalline phase of the generated zeolite does not depend on the kind microwave irradiation. The reason behind this statement is the time sample preparation without any irradiation was take longer than modified. This experimental result support the hypothesis that the microwave irradiation contributes to high crystallinity in short time.

Effect of microwave irradiation in purification process

Effect of zeolite adsorbent capability coupled with microwave irradiation is studied by adsorption process of waste cooking oil, which is, has 9 meq/kg of peroxide value. The peroxide value is fixed as parameter waste cooking oil quality. Process of adsorption is carried out by constant stirring and 60°C of heating in ratio 1:10 zeolite and waste cooking oil. The number of peroxide value each sample was analyzed by iodometric method. Result of peroxide value after adsorption process shows in table 2 as follows.

Table 2. The Peroxide Value before and after adsorption.

Sample	Peroxide Value (meq/kg)
Waste cooking oil	9
Unmodified zeolite	6,4
100 W adsorption	5
180 W adsorption	2

Microwave irradiation contribute in adsorption capability, it is indicated by decreasing the number of peroxide of wasted cooking oil. Increasing of annealing temperature will affect to the zeolite morphology i.e. surface area. Pakpahan and Co-workers (2017) [19] have been reported that increasing of temperature lead the decreasing of surface area and increasing particle size of zeolite. This morphology was contribute to the adsorption capability.

4. Conclusion

Zeolite with high crystallinity was success prepared by microwave irradiation. Comparing with conventional method, microwave irradiation increasing crystallinity of zeolite synthetic in short time (1 h). Unmodified sample (without irradiation) needs more preparation time (5 h). Microwave heating more effective in zeolite synthetic of rice husk ash than conventional heating. Microwave irradiation also increase the capability of zeolite synthetic applied in waste cooking oil purification. The peroxide value for waste cooking oil was decreased by adsorption with modified zeolite.

5. Reference

[1] Badan Pusat Statistik 01 March 2016 Accessed 2017 from *https://www.bps.go.id/brs/view/1271*
[2] Prasertsan S and Sajjakulnukit B 2006 Biomass and biogas energy in Thailand: Potential, opportunity and barriers *Renewable Energy* **31** pp 599-610
[3] Chandra S 1996 *Waste Materials Used in Concrete Manufacturing* (Westwood: William Andrew)
[4] Real C, Alcala M and Criado J.M. 1996 Preparation of silica from rice Husks *Journal of the American Ceramic Society* **79** pp 2012–16
[5] Patel M, Karera A and Prasanna P 1987 Effect of thermal and chemical treatment on carbon and silica contents in rice husk *Journal of Material Science* **22** pp 2457-64
[6] Kunihiro Fukui, Keiji Kanayama, Tetsuya Yamamoto and Hideto Yoshida 2007 Effects of microwave irradiation on the crystalline phase of zeolite synthesized from fly ash by hydrothermal treatment *Advanced Powder Technol* **18** pp. 381–393
[7] Yanagida S, Nikawa Y, Katsuki H, Kato K, Takizawa H and Tokuda M 2003 *New Industrial Technology of Microwave* pp. 4–8 (Japan: NTS)
[8] Somiya S and Roy R 2000 Hydrothermal synthesis of fine oxide powders *Bull. Mater. Sci.* **23** pp 453–460
[9] Querol X, Alastuey A, López-Soler A, Plana F, Andrés JM and Juan R. 1997 A fast method for recycling fly ash:microwave-assisted zeolite synthesis *Environ. Sci. Technol.* **31** pp 2527-33.
[10] Inada M, Tsujimoto H and Eguchi Y, Enomoto N and Hojo J 2005 Microwave-assisted zeolite synthesis from coal fly ash in hydrothermal process *Fuel* **84** pp 1482–6
[11] Arafat A, Jansen J C, Ebaid A R and Van Bekkum H 1993 Microwave preparation of zeolite Y and ZSM-5 *Zeolites* **13** pp 162-5

[12] Ponsot I, Bernardo E, Bontempi E, Depero L E, Detsch R, Chinnam R K and Boccaccini A R 2015 Recycling of pre-stabilized municipal waste incinerator fly ash and soda-lime glass into sintered glass-ceramics *J. Clean. Prod.* **89** pp 224-230

[13] Muljiyanti D R, Nuryono and Kunarti E S 2010 Sintesis dan karakterisasi silika gel dari abu sekam padi yang dimobilisasi dengan 3-(trimetoksisilil)-1 propantiol (Yogyakarta: Sekolah Pasca Sarjana Universitas Gadjah Mada Yogyakarta)

[14] Wahyuni T E 2003 *Hand Out Metode Difraksi Sinar X* (Yogyakarta: Laboratorium Kimia Analitik FMIPA UGM)

[15] Suryani P E, Candra A D 2017 Pengaruh Iradiasi Microwave Terhadap Luas Permukaan Spesifik Pada Zeolit Sintetis Dari Abu Sekam Padi *Seminar Nasional Hasil Penelitian (SNHP)-VII* Semarang

[16] Nair D G, Fraaij A, Klaassen A A K and Kentgens A P M 2008 A structural investigation relating to the pozzolanic activity of rice husk ashes *Cement and Concrete Research* **38** pp 861–869

[17] Auerbach S M, Carrado K A and Dutta P K 2003 *Hand book of zeolite science and technology Marcel Dekker* (New York: Basel,Switzerland)

[18] Fuadi A M, Malik Musthofa, Kun Harismah, Haryanto and Nur Hidayati 2012 Pembuatan Zeolit Sintetis Dari Sekam Padi *Simposium Nasional RAPI XI FT UMS-2K012*

[19] Pakpahan, Jennifer Kapriati, Karo, Pulung Karo, Suroto and Bambang Joko 2017 Studi Luas Permukaan Spesifik Zeolit Akibat Pengaruh Mikrostruktur dan Potensinya Sebagai Elektrode Superkapasitor *Jurnal Teori dan Aplikasi Fisika* **5** 1 pp 19-24

IOP Conf. Series: Materials Science and Engineering **395** (2018) 012002 doi:10.1088/1757-899X/395/1/012002

X-ray Absorption Studies on Cu–doped ZnO Nanoparticles Synthesized by Coprecipitation Method

D A Daratika[1], C Saiyasombat[2], K Kamonsuangkasem[2], M A Baqiya[1], and Darminto[1*]

[1]Department of Physics, Institut Teknologi Sepuluh Nopember (ITS), ITS Campus Keputih-Sukolilo, Surabaya 60111, Indonesia.
[2]Synchrotron Light Research Institute, 111 University Avenue, Muang District, Nakhon Ratchasima 3000, Thailand.

[*]Email : darminto@physics.its.ac.id

Abstract. Copper–doped ZnO (CZO) nanoparticles have successfully been synthesized by coprecipitation method with Cu concentrations ranging from 0 to 5 wt%. The effect of Cu concentration on electrical properties and the oxidation state of ZnO were investigated. The x-ray diffraction (XRD) pattern analysis shows CZO phase formed with calcination at 400°C for 3 h. The crystal size of CZO nanoparticles, evaluated using *MAUD* software, is in the range of 56 – 90 nm. It was obtained that the Cu doping has reduced the optical band gap from 2.85 eV to 3.06 eV, while the electrical conductivity increased from 1.03 x 10^{-8} to 24.25 x 10^{-8} S/cm. The highest conductivity of CZO is achieved by the sample doped with 4 wt% Cu. This result has been further studied by the x-ray absorption near edge spectroscopy (XANES) measurement at Cu *K*-edge related to the oxidation state of the Cu ions substituting Zn sites.

1. Introduction

Zinc oxide (ZnO) is a transparent semiconductor oxide with a wide direct band gap (3.37 eV), a large exciton binding energy (60 meV) at room temperature, and chemically and thermally stable. Due to these properties, it has emerged as a promising material for gas sensor, memory of optical and magnetic devices, piezoelectric transducers, photodiodes, photodetectors, UV-light emitting diodes (UV-LEDs), biomedical, solar cell, and transparent conductive oxide. Nowadays, synthesis of ZnO nanoparticles can be done by using several methods including a ball milling, auto-combustion, sol-gel process, hydrothermal route, and co-precipitation method [1].

Various efforts have already been made to investigate the effect of doping transition metals like Cu, Co, Al, Li, Ni, Fe, etc. on its electrical, optical, and magnetic properties. Among these doping elements, copper (Cu) is a potential dopant to improve the electrical conductivity which has low resistivity ($1.67x10^{-6}\ \Omega\ cm$), and has many physical and chemical properties that are similar to zinc (Zn), such as have a similar structure of the electron shells [2]. The presence of copper is used as a variable electrical conductance, thus it can control the electrical conductivity of ZnO. The Cu doping in ZnO changes the electrical properties but the effect of Cu in ZnO is debatable. There are many studies which have led to a number of hypotheses. Singhal, et al. (2012) have reported that Cu-doped ZnO (CZO) nanoparticles have Cu valence state of 2+ [2]. Suja et al. (2015) have found that CZO thin films have Cu valence state of 1+ [3]. Chakraborti et al. (2007) have reported that CZO thin films have mixed valence state of both Cu^{1+} and Cu^{2+}, which is also suggested that besides Cu^{2+} has substituted, the Cu^{1+} lies in the interstitial sites [4]. The XANES is one of the x-ray absorption spectroscopy (XAS) spectra region dominated by strong photoelectron scattering whose energy is located in the range of 40 eV above an absorption edge, and provides local information on materials such as oxidation state [5]. Sung et al. (2013) have reported the oxidation state of Cu in CZO thin films is 2+ determined by XANES measurements [6].

In the previous work, synthesis of CZO nanoparticles by coprecipitation has obtained that the Cu doping has reduces the optical band gap from 3.06 eV to 2.85 eV, while the electrical conductivity increased from 1.03 x 10^{-8} to 24.25 x 10^{-8} S/cm. The highest conductivity was achieved by the sample doped with 4 wt% Cu [7]. In this research, it will be studied further on this electrical conductivity of the CZO nanoparticles based on the oxidation state where Cu ions substituting Zn ions using XANES measurement at Cu *K*-edge.

2. Experimental Procedure

CZO nanoparticles have been synthesized by coprecipitation method with Cu concentrations ranging from 0 to 5 wt% (which is marked as CZO0 for the undoped ZnO, and CZO1, CZO2, CZO3, CZO4, CZO5, respectively) using zinc acetate dehydrate [$Zn(CH_3COO)_2.2H_2O$], and copper powder in certain compositions as raw materials, which were mixed and dissolved in an appropriate amount of 0.5 M HCl. The 0.5 M NH_4OH was gradually added to the solution by titration method to achieve pH around 9. The experimental procedure was similar with the previous work [7]. The crystal structure and phase of the samples were investigated using XRD with scanning angle 2θ ranging from 20° to 70° and operated at 40 kV and 30 mA with $CuK\alpha_1$ radiation with wavelength of 1.54 Å. The oxidation state where Cu has substituted Zn ions were investigated using XAS technique including XANES measurements at Cu *K*-edge, performed using synchrotron radiation at Synchrotron Light Research Institute (SLRI) using the beamline BL1.1W (Multiple X-ray Techniques) in Thailand.

3. Results and discussion

The phase purity and crystal structure of CZO nanoparticles for various doping concentrations have been studied by XRD patterns. Figure 1 shows the patterns of CZO nanoparticles. Diffraction peaks have been observed to be sharp implying high crystallinity of the samples. All samples show only reflection planes of ZnO phase indicating that Cu ions are dissolved into ZnO structure, so that Cu have substituted Zn ions without forming any impurity phases such as Cu, Cu_2O, and CuO. The XRD peaks show that CZO nanoparticles have wurtzite structure for all Cu concentrations.

The diffraction patterns were further analyzed by *Rietica* and *MAUD* (Material Analysis Using Diffraction) software using Rietveld method to evaluate the lattice parameter and crystallite size of the samples, respectively. The results of lattice parameters and crystallite size of CZO samples are tabulated in Table 1. Both of the lattice parameters and crystallite size decrease with the increasing concentrations of Cu. This indicates that Zn^{2+} (0.74 Å) ions are successfully replaced by the Cu ions.

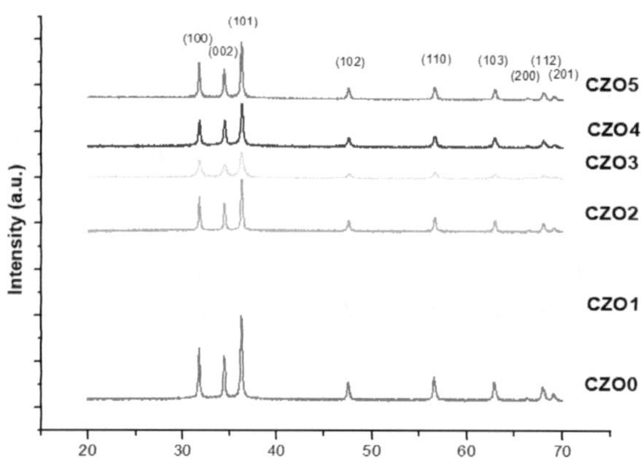

Figure 1. XRD patterns of Cu-doped ZnO (CZO) nanoparticles.

IOP Conf. Series: Materials Science and Engineering **395** (2018) 012002 doi:10.1088/1757-899X/395/1/012002

This result leads us for further investigation by XAS measurements at Cu *K*-edge to understand the information of Cu ions substituting the Zn sites. In the previous work, the Cu doping had been found to reduce the optical band gap from 2.85 to 3.06 eV, while the electrical conductivity increased from 1.03 x 10^{-8} to 24.25 x 10^{-8} S/cm. The highest conductivity was achieved by the sample with 4 wt% Cu doping [7]. Based on these results, it was expected due to the change of oxidation state of Cu ions and it may explain the highest conductivity phenomenon occurred in the sample with 4 wt% Cu doping.

Table 1. Crystallite size, and lattice parameters of Cu-doped ZnO (CZO) nanoparticles.

Copper content (wt%)	Crystallite size (nm)	Lattice Parameter				
		a = b (Å)	c (Å)	α = β (⁰)	γ (⁰)	Volume Cell
0	90	3.207590	5.135677	90	120	45.760006
1	84	3.207813	5.134898	90	120	45.759407
2	76	3.206976	5.134461	90	120	45.731655
3	71	3.206475	5.132681	90	120	45.701523
4	62	3.205377	5.133203	90	120	45.674858
5	56	3.205587	5.131305	90	120	45.663967

One of the most appropriate ways to determine the oxidation state of dopant atom is XANES measurement, analyzed by the *Athena* program. The XANES spectra at Cu *K*-edge for CZO nanoparticles (CZO3, CZO4, and CZO5) are shown in Figure 2 (a), and their derivatives in Figure 2 (b) along with the standard reference samples, which are Cu foil, Cu_2O, and CuO with *0*, *+1*, *+2* oxidation states of Cu, respectively. The different concentrations of Cu affected the XANES spectra.

The most primary region in the XANES Cu-K edge spectra is absorption edge. The edge region is formed by the transition of electrons from the deep core to the unoccupied outer shell depending on the absorber atom used in XANES measurements. The transition of the edge for Cu *K*-edge is from 1s to 3d. The electron configuration of Cu^{1+} is [Ar] $3d^{10}\,4s^0$, and that of Cu^{2+} is [Ar] $3d^9\,4s^0$, while Cu^{3+} is [Ar] $3d^8\,4s^0$. The transition of electrons from the deep core on the *K*-edge are from the 1s orbital to the unoccupied orbitals. However, Cu^0 and Cu^{1+} valence state does not have a hole in the 3d orbitals, so in this state, the transition of 1s to 3d is not allowed [6].

Figure 2 (a) shows the XANES Cu-K edge spectra of CZO nanoparticles (CZO3, CZO4, and CZO5) compared with the standard samples (Cu foil, Cu_2O, and CuO). The absorption edge energy (E_0) can be seen at the first peak of the XANES spectra which can be seen clearly using the first derivative in Figure 2 (b). The absorption edge energy positions of Cu reference were determined to be 8979 eV, 8980.14 eV, and 8983.58 eV for Cu foil, Cu_2O, and CuO, respectively. The absorption edge energy shifts to a higher value when the oxidation state increases. The edge positions of Cu were at 8985.01 eV, 8985.12 eV, and 8984.71 eV for CZO3, CZO4, and, CZO5, respectively. The absorption edge position of CZO4 has the highest edge energy. The greater value of edge energy indicates more photons absorbed by Cu (as an absorber atom) which is characterized by the high absorption edge peak.

Previous research carried out by Kataoka et al. (2011) described that the CZO nanowires had mixed valence state of both Cu^{2+} and Cu^{3+}, where Cu^{3+} ions were dominant in the surface area [8]. Deng et al. (2016) showed that the Cu^{3+} in $NaCuO_2$ material had been measured using XANES with an edge energy position of around 8984 eV [9]. Based on the comparison between the CZO samples and Cu standard samples, the oxidation state of all samples was approximately 3+. This is due to the oxidation occurred in the samples during synthesizing the nanoparticles at relatively high temperature, and also because of high Cu concentration as the dopant.

IOP Conf. Series: Materials Science and Engineering **395** (2018) 012002 doi:10.1088/1757-899X/395/1/012002

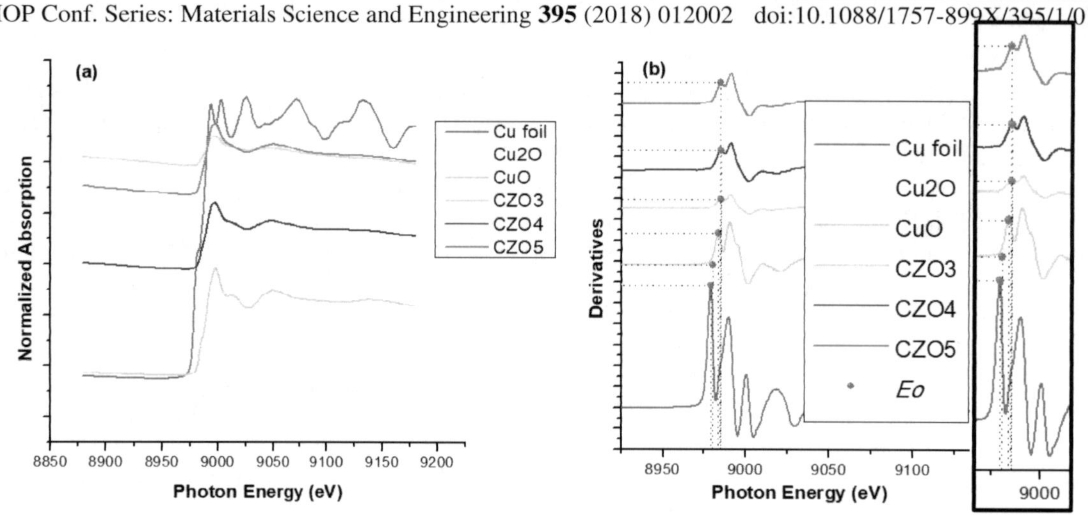

Figure 2. (a) Cu *K*-edge XANES spectra, and (b) their first derivatives of Cu-doped ZnO (CZO) nanoparticles.

Based on the present results, it can be seen that the decreases of the lattice parameters and crystallite size of CZO nanoparticles were caused by Cu^{3+} ions successfully substituting on Zn^{2+} sites. It is caused by Cu^{3+} ions having the ionic radius less than that of Cu^{2+} ions (0.73 Å), since Cu^{3+} ion loses one electron compared with "normal" Cu^{2+} ion. The CZO with concentration of 4 wt% Cu (CZO4 sample) has the highest electrical conductivity [7]. It is probably because the CZO4 sample has the highest absorption edge energy so that the oxidation state is higher than other samples affecting the amount of charge carrier in the sample. Thus, it may reduce slightly the band gap of CZO nanoparticles.

4. Conclusion

The CZO nanoparticles have been successfully synthesized using coprecipitation method. XRD characterization shows that CZO nanoparticles have single phase of wurtzite structure with crystal size in the range of 56 – 90 nm. The decrease of the lattice parameters and crystallite size of CZO nanoparticles was caused by Cu^{3+} ions successfully substituting on the position of Zn^{2+} sites. The oxidation states of Cu in all samples, determined by XANES measurements, were around 3+. The CZO4 sample has the highest electrical conductivity. It is probably due to the highest absorption edge energy of CZO4 so that the oxidation state is higher than the other samples providing more electronic charge carriers.

5. References

[1] Al-Ghamdi A A, Al-Hartomy O A, El Okr M, Nawar A M, El-Gazzar S, El-Tantawy F, Yakuphanoglu F 2014 *Spectrochim. Acta. A. Mol. Biomol. Spectrosc.* **131** 512
[2] Singhal S, Kaur J, Namgyal T, Sharma R 2012 *Phys. B. Condens. Matter* **407** 1223
[3] Suja M, Bashar S B, Morshed M M, Liu J 2015 *ACS Appl. Mater. Interfaces* **7** 8894
[4] Chakraborti D, Narayan J, Prater J T 2007 *Appl. Phys. Lett.* **90** 062504
[5] Ankudinov A L, Ravel B, Rehr J J, Conradson S D 1998 *Phys. Rev. B* **58** 7565
[6] Sung N-E, Kang S-W, Shin H-J, Lee H-K, Lee I-J 2013 *Thin Solid Films* **547** 285
[7] Daratika D A, Baqiya M A, Darminto 2017 *IOP Conf. Ser. Mater. Sci. Eng.* **196** 012009
[8] Kataoka T, Yamazaki Y, Singh V R, Fujimori A, Chang F-H, Lin H-J, Huang D J, Chen C T, Xing G Z, Seo J W 2011 *Phys. Rev. B* **84** 153203
[9] Deng Y, Handoko A D, Du Y, Xi S, Yeo B S 2016 *ACS Catal.* **6** 2473

Acknowledgment
This research was partially supported by "Hibah KLN", Kemenristekdikti, Indonesia, 2017**.** The use of experimental facilities at Synchrotron Light Research Institute (SLRI), using beamline BL1.1W (Multiple X-ray Techniques), Thailand, was thankfully appreciated.

IOP Conf. Series: Materials Science and Engineering **395** (2018) 012003 doi:10.1088/1757-899X/395/1/012003

Effect of Substrate Orientation on the Growth Direction of $In_xGa_{1-x}As$ Nanowires (NWs)

E Wibowo[1]*, N Ulya[1], Z Othaman[2], P Marwoto[3], I Sumpono[3], M P Aji[3], Sulhadi[3], B Astuti[3], M Rokhmat[1], Suwandi[1], A Ismardi[1] and Sutisna[4]

[1]Engineering Physics, School of Electrical Engineering, Telkom University, Jalan Telekomunikasi No.1, Terusan Buah Batu, Bandung, Indonesia.
[2]Department of Physics, Faculty of Sciences, Universiti Teknologi Malaysia (UTM), Skudai, Johor Bahru, Malaysia.
[3]Department of Physics, Faculty of Mathematics and Natural Sciences, Universitas Negeri Semarang (UNNES),Gunungpati, Semarang, Indonesia.
[4]Department of Physics, Faculty of Mathematics and Natural Sciences, Universitas Negeri Jember, East Java, Indonesia.

* Email: edywibowo86@yahoo.com; edywibowo@telkomuniversity.ac.id

Abstract. We have grown the $In_xGa_{1-x}As$ NWs on GaAs(111), GaAs(100) and Si(111) substrates via Vapor-Solid-Solid (VSS) mode using MOCVD. We observed that the cylindrical NWs grow perpendicular to the GaAs(111) substrate. The straight line NWs with an angle of 50.60° to the normal were occurred on GaAs(100), while kinks NWs were perceived on Si(111). We found that the growth direction of the $In_xGa_{1-x}As$ NWs can be easily controlled using certain orientation of the substrate by considering its lattice mismatch and its surface energy.

1. Introduction

Semiconductor NWs have been continuously gaining interest among science, researcher and engineer communities since the late 1990s, pioneered by Wagner and Ellis [1]. The vapor liquid solid VLS is one of the most common model to grow NWs assisted by seed particles [2-5]. In this model vapor donates the phase of precursors, liquid the phase of particles, and solid the phase of NWs. Nevertheless, it has been reported that the particle could also be in a solid state when growth occurs, which is labelling as the vapor solid solid (VSS) model [4-15]. In particular, III-V nanowires can be epitaxially grown by metalorganic chemical vapor deposition (MOCVD) [2, 3]. In the NWs growth process using MOCVD, nanoparticle as a seed promotes NWs growth by either VLS or VSS mechanism [16, 17]. At growth temperature, the seed particles on the semiconductor substrate surface form a liquid (for VLS) or solid (for VSS) alloy with the group III species [2-5, 17]. Once the group III species precipitate out at the seed particle-semiconductor interface, nanowire growth occurs with seed nanoparticle a top each nanowires [1-10, 17]. In this paper, the effect of substrate materials and it's orientation on the growth direction of gold seed-particles assisted growth of $In_xGa_{1-x}As$ NWs are deeply investigated.

2. Experiment

The $In_xGa_{1-x}As$ NWs were grown by Au seed particles-assisted on GaAs(111)B and (100) as well as Si(111) substrates using MOCVD. The system uses low-pressure (0.1 atm) chamber with trimethylindium (TMIn), trimethylgallium (TMGa) and arsine (AsH_3) as precursors. Palladium-diffused hydrogen (99.9999% pure hydrogen) is used as carrier gas. The TMGa and TMIn are held in temperature controlled baths at -3.5 °C and 17 °C, respectively, while arsine is diluted to 10% in hydrogen. The temperature and pressure of the metal-organic precursors are fixed to control the vapor pressure in the bubbler. The substrates were functionalized by immersed it in 0.1 % poly-L-lisine (PLL) solution for 3 minutes. Then, it was cleaned with de-ionized water and subsequently dried with nitrogen (N_2). It was then treated with 50 % gold colloid solution for 30 seconds. The negatively charged of Au particles are attracted to the positively charged PLL layer thus are immobilized on the substrate surface [13, 14].

After Au seed particles were deposited on substrate surface, the substrate was placed in MOCVD chamber. Substrate was heated at temperature of 600 °C under a constant partial pressure of AsH_3 gas for 10 minutes, and then cooled to desired growth temperature. Once the growth temperature was achieved, TMIn, TMGa, and AsH_3 were flowed into the chamber and the growth begun. The growth time, growth temperature and V/III ratio were set at 30 minutes, 400 °C and 10, respectively. The growth temperature is below the pseudobinary eutectic point Au-GaAs (630 °C) [15]. Therefore, the state of gold seed particle is solid or molten; its surface and interface are liquid, while the core of the seed particle is solid [16]. For that reason, $In_xGa_{1-x}As$ NWs were grown via VSS rather than VLS mechanism. The samples were then investigated using Field Emission Scanning Electron Microscopy (FESEM) located at Ibnu Sina Institute for Fundamental Science Studies, Universiti Teknologi Malaysia to investigate the morphology of the NWs.

3. Results and discussion

The effects of substrate orientation on the growth direction of $In_xGa_{1-x}As$ NWs have been deeply investigated in this section. $In_xGa_{1-x}As$ NWs were grown on GaAs(111) and (100) as well as on Si(111) substrates with similar growth parameters. The growth temperature was 400 °C, while the V/III ratio and (x) were 10.0 and 0.47, respectively. FESEM images of $In_xGa_{1-x}As$ NWs grown with different substrate orientation is shown in Figure 1.

Figure 1(a and b) show the $In_xGa_{1-x}As$ NWs grown on GaAs(111) grow perpendicular to the surface of substrate. The direction of $In_xGa_{1-x}As$ NWs can be determined using the relationship of crystallography plane and crystallography direction. For the cubic system, if the growth direction is perpendicular to an (h k l) plane then the orientation of the crystal is an [h k l] direction [17, 18]. It means the direction of $In_xGa_{1-x}As$ NWs grow perpendicular to the GaAs(111) substrate is [111]. Nonetheless, $In_xGa_{1-x}As$ NWs grown on GaAs (100) substrate did not grow perpendicular to the substrate but grow straight line form an angle of 50.60^o to the normal.

By using the relationship of crystallography plane and crystallography direction, the growth direction of $In_xGa_{1-x}As$ NWs can be also identified i.e. [111]. This angle undergoes deviation of 4.14^o from the theoretical reference [21]. On the reference, the angle between crystallographic plane of (100) and crystallographic direction of [111] should be 54.74^o. The deviation of this value is likely contributed by the miss cut of GaAs(100) substrate. In addition, it can be obviously observed that the NWs are grown both on GaA(111) and (100) substrate grow in the same orientation to [111] direction. This indicates that the [111] is the preferential growth direction of NW as its lower surface energy [18, 19]. Consequently, NWs grown on (111) grow perpendicular to the substrate while the NWs grown on GaAs(100) grow to form an angle of 50.60^o to the substrate. Nevertheless, once the $In_xGa_{1-x}As$ NWs grown on Si(111) substrate, NWs did not grow upright as grown on GaAs(111) and it shape is not elongated, although the crystallography plane of Si and GaAs substrates is similar. This phenomena is assumed due to the surface energy differences between GaAs(111) and Si(111) as well as the lattice mismatch differences between $In_xGa_{1-x}As$ NWs and substrate. In practice, the surface energy of any surface material can be calculated by using Eq.1 as following:

IOP Conf. Series: Materials Science and Engineering **395** (2018) 012003 doi:10.1088/1757-899X/395/1/012003

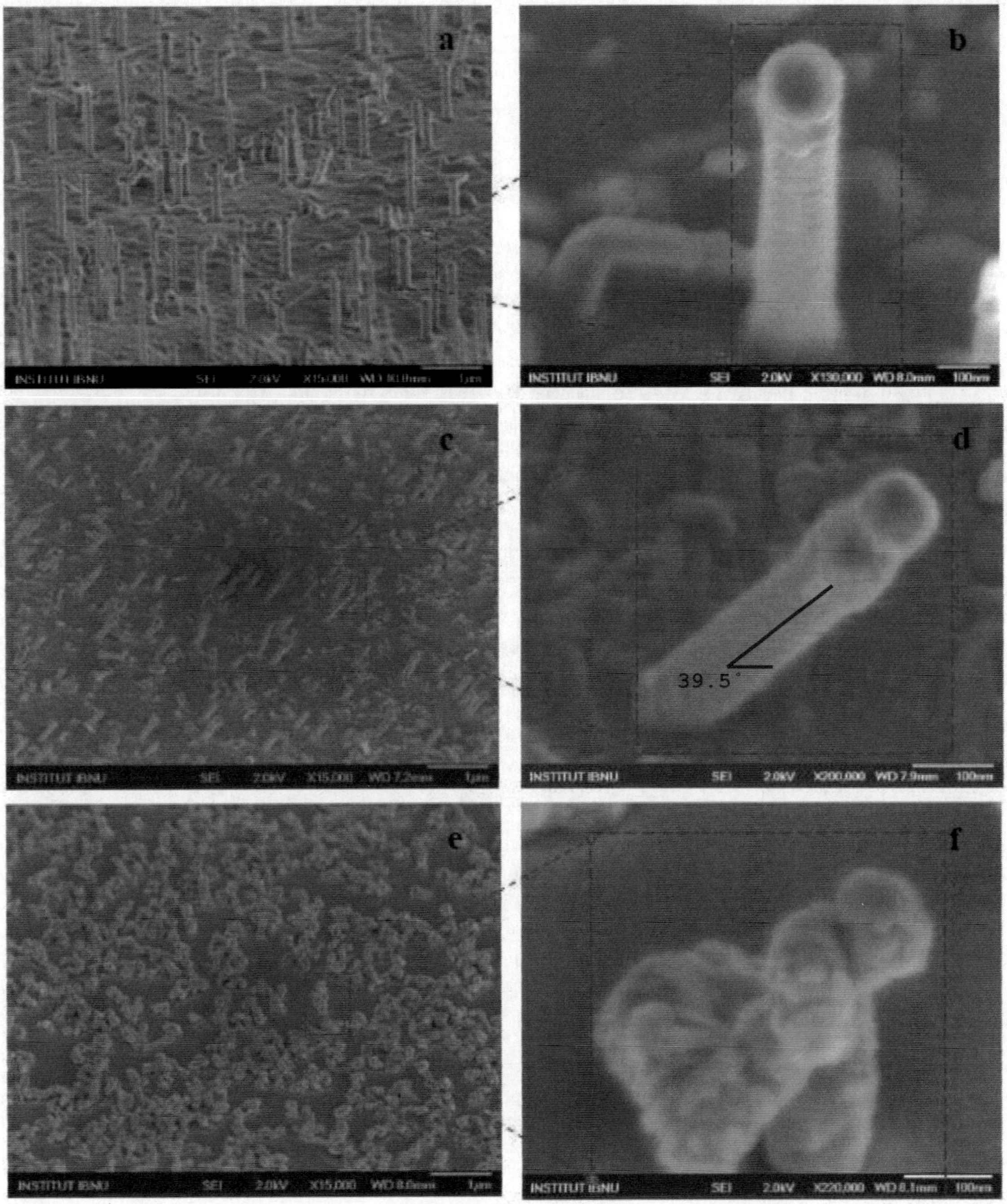

Figure 1. FESEM images of $In_xGa_{1-x}As$ NWs grown with different substrate orientation: (a, b). on GaAs (111)B; (c, d). on GaAs (100)B and (e, f). on Si (111).

$$\gamma = \frac{1}{2} N_b \varepsilon \rho_a \tag{1}$$

where γ is the surface energy, N_b is the number of broken bonds, ε is the bond strength and ρ_a is the surface atomic density i.e. the number of atoms per unit area on the surface [19]. *Due* to the crystal structure of GaAs is zinc blende and the crystal structure of Si is diamond, both of this material has similarity on the number of atoms and the crystallography geometry. Therefore the number of broken bonds of this material (N_b) on the (111) plane is assumed equal. Moreover, based on the definition of the surface atomic density, the value of ρ_a is inversely proportional to the square of the lattice constant (a). Hence to simplify the calculation, the surface energy (γ) can be then expressed as function of (ε) and (a) as follow:

$$\gamma \approx \frac{\varepsilon}{a^2} \quad (2)$$

Based on Eq. 2 it can be seen that the surface energy of material highly dependable on the bond strength and the lattice constant of material. In case of GaAs and Si substrate, the bond strength of GaAs(111) is lower than Si(111) [20], whereas the lattice constant of GaAs (5.6534 Å) is higher than Si (5.4307 Å). Therefore, the surface energy of Si(111) is higher than GaAs(111), even though the crystallography plane both of those material is identical. Consequently, $In_xGa_{1-x}As$ NWs grown on Si(111) substrate did not grow straight as NWs grown on GaAs(111) substrate. Adachi has also reported that the surface energy of GaAs(111) and Si(111) i.e. 1.05 Jm^{-2} and 1.36 Jm^{-2}, respectively [20]. Besides, the difference of lattice mismatch between $In_xGa_{1-x}As$ NWs to the substrate is another aspect that considered could contribute on the difficulty level of growth NWs on certain substrate. The range of the lattice constant of ternary alloy $In_xGa_{1-x}As$ is from 5.6534Å (GaAs) to 6.0585Å (InAs). It means the lattice constant of Si substrate is out of the range from the lattice constant of $In_xGa_{1-x}As$, while the lattice constant of GaAs substrate still inside this range. As a result, the lattice mismatch $In_xGa_{1-x}As$ NWs to the Si substrate is higher than to the GaAs substrate. Therefore, $In_xGa_{1-x}As$ NWs are easier grown on GaAs rather than on Si substrate due to its lower surface energy and lattice mismatch. Consequently, $In_xGa_{1-x}As$ NWs grow straight on GaAs(111) substrate but grow with kinks morphology on Si(111) substrate. So, the kinds and orientation of substrate are important aspect should be considered in order to epitaxially growth of $In_xGa_{1-x}As$ NWs. As well, the growth direction of NWs can be possibly controlled using certain orientation of the substrate by considering its lattice mismatch and surface energy.

4. Conclusion

The investigation of the effect of substrate orientation on the growth direction of $In_xGa_{1-x}As$ NWs were conducted. It can be seen that the growth direction of NWs is possible controlled. Cylindrical $In_xGa_{1-x}As$ NWs grow perpendicular to the substrate along [111] direction when it is grown on GaAs(111). On GaAs(100), the NW's orientation is straight line form an angle of 50.60° to the normal on the [111] direction, whereas kinks NWs are observed on Si(111). The growth direction of NWs can be controlled using certain orientation of the substrate by considering its lattice mismatch and surface energy.

Acknowledgment

We would like to thank Ministry of Science, Technology and Innovation Malaysia & Ministry of High Education Malaysia for financial support via FRGS grant number 78339 and Ibnu Sina Institute for Fundamental Science Studies, Universiti Teknologi Malaysia for laboratory facilities.

References

[1] Wagner R S and Ellis W C 1964 *Appl. Phys. Letters* **4** 89
[2] Wibowo E, Othaman Z, Sakrani S, and Sumpono I 2011 *Nano* **6** 159
[3] D Gustiono D, Wibowo E, and Othaman Z 2013 *J. Phys.: Conf. Ser.* **423** 012047
[4] Wibowo E, Othaman Z, Sakrani S, Ameruddin A S, Aryanto D, Muhammad R and Sumpono I 2011 *Journal of Applied Sciencess* **11** 1315
[5] Wibowo E, Othaman Z and Sakrani S 2013 *Advanced Materials Research* **667** 224
[6] Zaraska L, Mika K, Syrek K, and Sulka G D 2017 *Journal of Electroanalytical Chemistry* **801** 511
[7] Ali L M and Abed F A 2017 *Optical Materials* **72** 650

[8] Eun H R, Yoon Y J, Seo J H, Cho M S, Lee J-H, Kwon H-I and Kang I M 2016 *Current Applied Physics* **16** 681

[9] Li A, Zou J and Han X 2016 *Sci China Mater* **59**(1) 51

[10] Kumar M , Kumar V and Singh R 2017 *Scripta Materialia* **138** 75

[11] Fortuna S A and Li X 2010 *Semicond. Sci. Technol.* **25** 024005 (16pp)

[12] Du W, Yang X, Pan H, Wang X, Ji H, Luo S, Ji X, Wang Z and Yang T 2015 *Cryst. Growth Des.* **15** 2413

[13] Regolin I, Sufdeld D, Luttjohann S, Khorenko V, Prost W, Kastner J, Dumpich G, Meier C, Lorke A and Tegudej F J 2007 *Journal of Crystal Growth* **298** 607

[14] Xiangfeng D and Lieber C M 2000 *Adv. Mater.* **12** 4

[15] Wang N, Cai Y and Zhang R Q 2008 *Science and Enginering R* **60** 1

[16] Runyan W R and Shaffner T J 1998 Semiconductor measurements & Instrumentation (second edition). United stated of America: McGraw-Hill

[17] Suryanarayana C and Norton M G 1998 *X-Ray Deffraction A Practical Approach*. New York: Plenum Press

[18] Kim Y 2006 *Nano Letters* **6** 599

[19] G. Cao. Nanostructures & Nanomaterials (Synthesis, Properties & Applications. London: Imperial College Press, 2003

[20] Adachi S 1994 GaAs and Related Materials: Bulk Semiconducting and Superlattice Properties. Singapura: World Scientific

IOP Conf. Series: Materials Science and Engineering **395** (2018) 012004 doi:10.1088/1757-899X/395/1/012004

The Influence of Temperature and Gas Flow Rate on the Formation $CaCO_3$ Vaterite Phase

Z Arifin*, M Zainuri, Y Cahyono and Darminto

Advanced Materials Research Group, Department of Physics, Faculty of Science, Institut Teknologi Sepuluh Nopember (ITS), Kampus ITS Sukolilo, Surabaya 6011, Indonesia

*E-mail: zaenalw18@gmail.com

Abstract. $CaCO_3$ powder is much needed in the industry. $CaCO_3$ vaterite phase is widely used as raw material in the pharmaceutical and cosmetic industries. PCC can be synthesized from natural limestone. Indonesia has abundant limestone reserves and has a very high purity[1]. In this study, the synthesis of $CaCO_3$ powder was carried out at 29 °C and 40 °C by carbonation method without and by using gas bubble breakers. Characterization of $CaCO_3$ with SEM and XRD shows the formation of calcite and vaterite phases. The formation of the vaterite phase $CaCO_3$ is very sensitive in the changes of temperature and CO_2 gas flow rate. The formation of a vaterite phase without a gas bubble breaker is increased linearly. In contrast to the gas bubble breaker the formation of vaterite is saturated after a CO_2 gas flow rate of 2.3 lt/min.

1. Introduction

$CaCO_3$ is polymorphic because it has three different types of crystal phases, aragonite, calcite and vaterite. Calcite with rhombic morphology is the most stable phase and easy to formed crystalline phase. Vaterite with microsphere morphology is the most unstable phase and the most difficult phase to form. The formation of the vaterite phase is influenced by many parameters, such as pH, temperature, and reactant concentration [1].

Calcium carbonate ($CaCO_3$) has been widely used in various industrial fields, such as papers, plastics,coatings, ink, paint and polymer products. Meanwhile, higher and special quality $CaCO_3$ has also been needed for ingredients from cosmetics [2], drug delivery [3], bioactive materials [4], to supplements of nutrition [5]. The type of calcium carbonate precipitates is used in the industry. The precipitated $CaCO_3$ (PCC) is generally prepared by precipitation process using CO_2, gas bubbling in the solution containing Ca ions, such as $Ca(OH)_2$. The starting materials $Ca(OH)_2$ may potentially be a limestone as natural resources with calcium content of around 98% [6].

PCC with varying morphology are generally synthesized from $CaCl_2$ solutions, either by mixing methods or by carbonation. Indonesia has an abundant source of limestone. $CaCO_3$ powder is synthesized from natural limestone. In our previous study, the PCC has successfully been prepared from limestone employing a facile method of carbonation, resulting in $CaCO_3$ with various phases and uniqe morphology [7].

In this work, we have succeeded in synthesizing $CaCO_3$ vaterite and calcite phases by carbonation method. The $Ca(OH)_2$ precipitating reaction with CO_2 gas is carried out with and without gas bubble breakers with different CO_2 gas flow rates at 29 °C and 40 °C.

IOP Conf. Series: Materials Science and Engineering **395** (2018) 012004 doi:10.1088/1757-899X/395/1/012004

2. Material and Methods

2.1. Materials

The limestone from a quarry in Tuban (East Java, Indonesia) has a 98% calcite content, as the starting material, was calcined for 5 hours at 900 °C. 40 gram of calcined products dissolved into 20 liters of demineralized water for 20 hours. Furthermore a clear $Ca(OH)_2$ solution will be used for the synthesis of $CaCO_3$ powder.

2.2. Synthesis procedure

The synthesis of 1 liter of $Ca(OH)_2$ solution was carried out at 29 °C and 40 °C with 3 steps:

- Step 1: CO_2 gas is flowed through 0.4 cm diameter pipe into $Ca(OH)_2$ solution and the solution is stirred at 200 rpm rotation rate.
- Step 2: CO_2 gas is flowed through 0.4 cm diameter pipe at which the end of the pipe is subjected to a gas bubble breaker and inserted into a stirred $Ca(OH)_2$ solution at a rate of 200 rpm rotation.
- Step 3: CO_2 gas is flowed through 0.4 cm diameter pipe which at the end of the pipe is given a gas bubble breaker and inserted into solution of $Ca(OH)_2$ which is stirred at high rotation rate.

2.3. Characterization

The powder diffraction patterns were recorded using XRD (Philips X'Pert MPD system, monochromated Cu-Kα radiation), employing record in the diffraction angle (2θ) range of 20° – 70°. The morphology of powder was observed using SEM (FEI Inspect S50, at the acceleration voltage of 15 kV), by dropping the particle suspension on a glass slide, drying overnight, then sputtering with gold-palladium in a vacuum evaporator (Emitech SC7620 *sputter cooter*).

3. Results and Discussion

3.1. Effect of gas bubble breakers and stirring speed

The process of formation of carbonate ions from CO_2 gas occurs in the layer between the liquid-gas interface [8]. In the case of carbonation in $Ca(OH)_2$ solution, this layer is in the gas bubble wall while in the solution. The total surface area of the bubbles in the liquid is inversely proportional to its volume. If the surface area increases, then the number of carbonate ions formed per unit time becomes more. This opinion can be an explanation of why the use of bubble breakers can accelerate the carbonation process.

The results of the previous study [7] suggest that the dissolved $CaCO_3$ mass reaches the optimum conditions occurring at ph 7 - 8. The time required for the solution to reach pH 7-8 at different flow rates is shown in Figure 1.

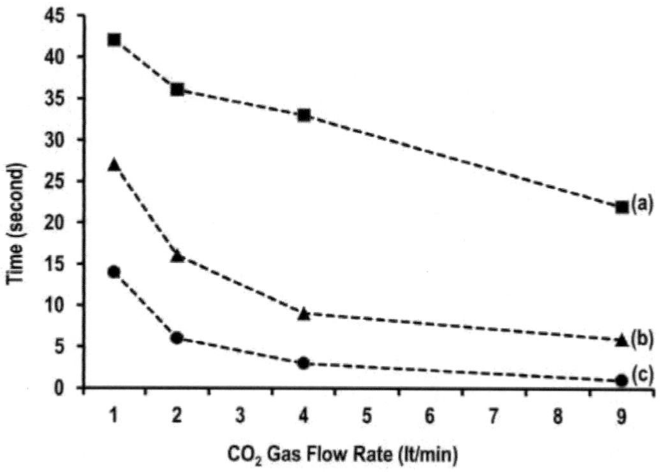

Figure 1. The time required for the solution to reach pH 7-8 at different CO_2 gas flow rates: (a) step 1 (b) step 2 (c) step 3

Next will be discussed saturation at flow rate above 2.35 lt/min (Figure 1). Factors affecting CO_2 gas absorption rate in solution are reaction rate, mass transfer, and solution concentration. The types of reactions that occur in this case are chemical and physical reactions. The physical reaction takes place when CO_2 with a diffused gas phase into the solution produces dissolved CO_2. While chemical reactions occur in the formation of various species of ions to the formation of $CaCO_3$, this chemical reaction takes place much shorter than the physical reaction. Thus, the saturation state of Figure 1b may be due to the low rate of solubility of the CO_2 gas and not due to the slow reaction of ion formation.

Rapid carbonation time is generated in synthesis using step 3 (Figure 1c). The increase of CO_2 gas flow rates can be explained from three aspects: the size of the gas bubble, the process of mass transfer of $CaCO_3$ from the bubble surface, and the chance of collision between the reactant particles.

The bubble size is determined from the buoyant force equilibrium against surface tension and bubble adhesive style with nozzle (bubble breaking pores). Stirring at high rotation gives an external force that contributes negatively to the last two forces, consequently bubbles can break away from nozzles with relatively smaller sizes (compared to step 2). As mentioned earlier, a small size of gas bubble can increases the total surface area of the liquid-gas interface.

The next reason is due to mass transfer of $CaCO_3$ as a reaction product. In chemical reactions, the higher concentration of the reactant the faster the reaction occurs, on the contrary, the product concentration will slow down the reaction rate. If the reaction of carbonic ion formation occurs in the liquid-gas interface then the reaction of $CaCO_3$ molecule formation also occurs in this area. In low agitation, the $CaCO_3$ (solid) molecule does not immediately leave the liquid-gas interface area. This, of course, slows down the formation of other $CaCO_3$ molecules. With high stirring, the $CaCO_3$ molecule will leave the liquid-gas interface area rapidly, so that the reactions to the gas bubble wall remain normal.

The last allegation is viewed from the aspect of chance of collision between reactant ions. The use of a blender is intended to force the reaction medium to move faster. The non-smooth wall structure of the tubes causes the fluid (reaction medium) to move turbulently. Although it does not increase the ionic energy significantly, it can increase the chance of collision rate accumulation which contributes to the increase of reaction rate.

IOP Conf. Series: Materials Science and Engineering **395** (2018) 012004 doi:10.1088/1757-899X/395/1/012004

3.2. Effect of variation of carbonation rate

The first treatment was performed to find the optimum synthesis temperature for the formation of vaterite. This is done by varying the solution temperature at 29 °C and 40 °C. Synthesis was carried out in standard solution with 200 rpm stirring and flow rate of 2.3 lt /min.

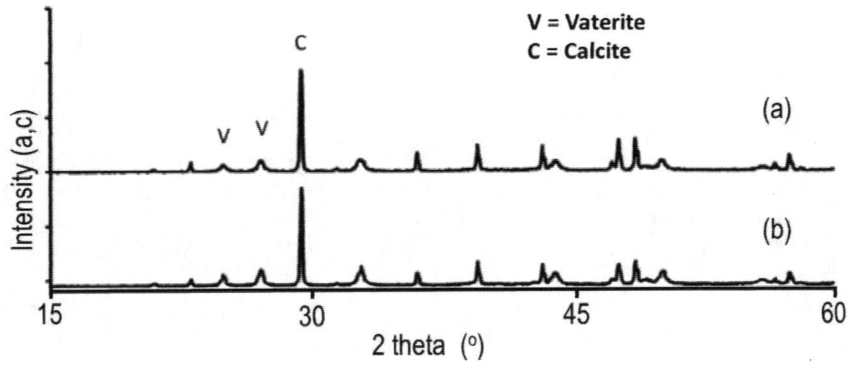

Figure 2. The XRD spectra of samples synthesized at temperature (a) 29 °C (b) 40 °C.

Based on XRD pattern formed calcite and vaterite phase. At both temperatures a slight increase in the vaterite fraction is characterized by a decrease in calcite peak height 104 in the angle $2\theta = 29°$. Further data processing showed changes in vaterite fraction from 43 wt% at 29 °C to 50 wt% at 40 °C. The results of this observation are consistent with the results of the study [9], although carried out with different materials and methods. Thus, the optimum temperature for the highest vaterite synthesis is 40 °C. The heavy fraction of vaterite refinement results with MAUD software is shown in Figure 3.

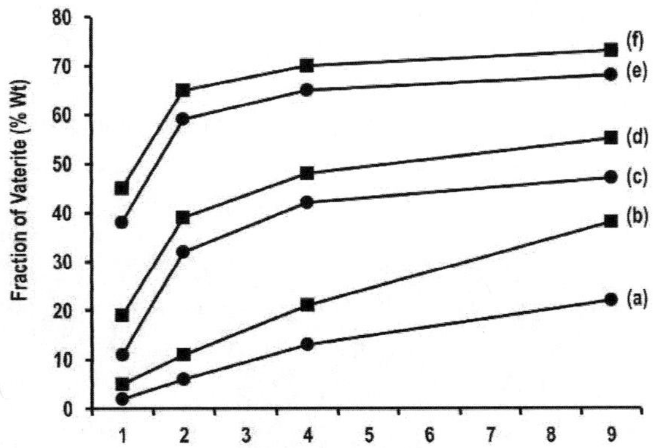

Figure 3. The resulting vaterite weight fraction at CO_2 gas flow rate and different carbonation: (a) T= 29 °C step 1 (b) T= 40 °C step 1 (c) T=29 °C step 2 (d) T= 40 °C step 2 (e) T= 29 °C step 3 (f) T=40 °C step 3

According to Figure 3d, the vaterite fraction is saturated at a fraction of about 57 wt% weight fraction, as well as its carbonation rate is also saturated. The veterite fraction at different step with CO_2 gas flow rate of 2.3 lt/min at 40 °C showed significant difference, 11 wt%, 53 wt%, and 65 wt%

respectively. From this data it is known that the increase of veterite weight fraction is agree with the rate of carbonation process.

Characterization with SEM samples synthesized by different steps is shown in Figure 4. The synthesis of $CaCO_3$ in step 1 yields rhombic-shaped particles with a calculated phase composition of 89 wt% (Figure 4a). Figure 4b is a sample synthesized with step 2 with a flow rate of 2.3 lt/min at 40 °C producing particles with rhombic and spheroidal morphology with 47 wt% and 53 wt% compositions. Figure 4c is a sample synthesized with step 3 producing particles dominated by speroidal vaterite (65 wt%) measuring about 2 μ.

Figure 4. SEM Photography sample synthesized with different carbonation rates:
(a) Step 1 (b) Step 2 (c) Step 3

The number of particles with the spheroidal morphology of Figure 4b appears dominant although the fraction weighs only 53%. The difference between these two characterizations is the impact of lower density of vaterite particles than calcite. Calcite is a solid particle, whereas vaterite is composed of nano-shaped spheroid aggregations to form a large porous spheroid. In this case, the effect of transformation also needs to be considered because of the difficulty of conditioning a truly objective synthesis treatment. low pH due to excessive carbonation reaction will increase the rate of transformation of vaterite into calcite.

The particle size formed from these three steps also looks varied. This difference is probably due to the stirring rate which disrupts the growth equilibrium of particle size. The next possibility is due to the difference in nucleation rate as the rate of reaction increases. The first possibility is easily denied because the samples in Figures 4a and 4b have different sizes although stirred at the same rate. In Figure 4c, the calcite rhombic particle appears to have an average size greater than that of the surrounding spheroidal particles, this is seen more clearly in Figure 5.

Figure 5. The calcite crystal growth process of diffused vaterite particles.

This situation is certainly interesting to explain further related to the process of crystal growth. One of the mechanisms of crystal growth is through the diffusion of soluble particles onto the surface of more stable particles. the porous area and its unstable thermodynamic properties, the vaterite acts as a soluble particle. Figure 5 shows some vaterite particles attached to the calcite rhombic surface. Vaterite will transform gradually until it blends into a calcite particle with a larger size. This mechanism is consistent with the results of experiment.

Two other mechanisms of $CaCO_3$ crystal growth were also observed in this study. The first mechanism occurs due to the aggregation of some stable crystalline particles to form new particles of a larger size [8], the visualization can be observed in Figure 5. The next mechanism is the gradual increase of crystal size due to the oriented aggregation of ions, molecules, or nanocrystals at surface of substrate [11].

3.3. Condition of vaterite and calcite formation

Synthesis with step 1 at a flow rate of 1.1 lt/min resulted in the amount of CO_3^{2-} formed in a relatively small amount of time interval. The high $[[Ca^{2+}] / [CO_3^{2-}]]$ ratio causes the amorphous molecules of $CaCO_3$ to tend to build a rhombohedral calcite structure (Fig. 4a). In the condition of such a reaction medium, the vaterite that had formed would be easy to transform. Conversely, at higher carbonation rates, the amount of CO_3^{2-} which can form within a certain time interval is considerable, so that the vaterite fraction increases, as in the synthesis results with step 2 and step 3.

The next study is the vaterite fraction which can not reach 100%, although it is synthesized with the maximum carbonation rate. This problem is thought to arise when the carbonation process is almost complete pH of the reaction medium begins to decrease as the amount of OH^- starts to decrease. The more dominant bicarbonate ion (HCO_3^{2-}) formed at low pH leads to high supersaturation conditions as the requirement for the formation of vaterite is not met. At low pH and supersaturation, vaterite is also more easily dissolved to further transform to calcite. Thus, synthesizing with carbonation methods on fixed volumes without involving additives would be extremely difficult to produce maximum vaterite.

4. Conclusion

Synthesis at low concentrations produces micron-sized particles with rhombic calcite and spheroidal vaterite morphology. Veterite with spheroid morphology tends to form at high carbonation rates. The maximum weight fraction of the synthesized vaterite $CaCO_3$ micro-sized powders is around 73 wt%, obtained from synthesis step 3 at 40 ^oC.

Acknowledgements

This research was partially supported by Dana Non PNPB ITS, Tahun 2017.

References

[1] Han Y S 2006 *J. Euro Ceram. So.* **26** 843
[2] Wang C, Tong H C, Liu Z and Zeng X F 2010 *Int. J. Pharmace.* **308** 160
[3] Peng C, Zhao Q and Gao C 2010 *Colloids Surf A: Physicochem Eng. Asp.* **353** 132
[4] Kumar G S 2008 *J. Colloid Interf. Sci.* **349** 56
[5] Ghamgui H 2007 *Biochem. Eng. J.* **37** 34
[6] Arifin Z, Pratapa S, Triwikantoro and Darminto 2013 *AIP Conf.Proc.* **1554** 90
[7] Arifin Z, Pratapa S, Triwikantoro and Darminto 2015 *Trans. Ind. Ceram. So.* **74** 1
[8] Judat B and Kind M 2004 *J. Colloid Interf. Sci.* **269** 341
[9] Ogino T and Sawada 1990 *J. Cryst Growth.* **100** 159
[10] Montes H, Renard F, Geoffroy N, Charlet and Pironon J 2007 *J. Cryst. Growth.* **308** 228
[11] Kedra, Krolik K and Gierycz P 2009 *J. Cryst. Growth.* **311** 3674

IOP Conf. Series: Materials Science and Engineering **395** (2018) 012005 doi:10.1088/1757-899X/395/1/012005

Graphene Oxide- Inorganic Composite Membrane: A Review

M Zunita[1,*], **I G B N Makertiharta**[1], **R Irawanti**[1], **N Prasetya**[2], **I G Wenten**[1]

[1] Department of Chemical Engineering, Institut Teknologi Bandung, Ganesha Street 10, 40132, Bandung, West Java, Indonesia
[2] Barrer Centre, Department of Chemical Engineering, Imperial College London, Exhibition Road, London, SW7 2 AZ, United Kingdom

*m.zunita@che.itb.ac.id

Abstract. Nowadays, membrane technology is one of excellent separation method. However, membranes as usual used have the heterogeneous pore size, poor mechanical property, thermally unstable, and fouling problem, therefore becomes a major challenge for many researchers. One of the most promising ways to address this issue is by making a composite membrane by introducing graphene oxide (GO). Graphene oxide (GO) is a two-dimensional material endowed with unique physicochemical, electrical, and thermal properties. Furthermore, GO has a large area and ultrathin layer surface, which is suitable for membrane technology as separation technology method. Its properties can be reducing volume area ratio of membrane to maximize flux and minimize transport resistance. Because of these exceptional properties, GO is then considered of being able to improve the separation performance of the native membranes. Recent developments in this area have shown the improvement of separation property coming from the GO-composite membrane. This is also followed by improvement in the physicochemical property of the resulting composite membrane such as in thermal and mechanical properties. In practical application, such GO-composite membranes have also been shown to be feasibly applied to separate heavy metal ions, mixture of gases, and organic materials. In addition, the GO-composite membrane is also suitable to be applied in fuel cell technology and has shown the antifouling and antibacterial property. This review paper will then focus on the recent developments of GO-composite membrane and factors affecting its successful synthesis so it can deliver exceptional separation performance.

1. Introduction

Membrane filtration is an advanced technology in fluid separation. Membrane technology offers several advantages such as operation simplicity, low footprint, energy intensive, intensified process, and low operating cost, which is made it a promising alternative to competing conventional technologies in almost industrial processes. Therefore, membrane has been widely adopted in various industrial processes including water treatment, wastewater treatment, chemical reaction, product purification, gas separation, and energy conversion [1,2].

In membrane-based processes, membrane material has to have mechanical integrity, thermal stability, and good selectivity, especially when encountering intense process condition. Furthermore, membrane fabrication also needs to consider the economic feasibility.

In order to address the above issues, graphene oxide (GO) is a promising material that has attracted numerous researchers in order to improve membrane properties and separation performance. [3–5]. This is because GO has exceptional physicochemical properties including mechanical resistance, thermal stability and chemically robust. Such a property is then beneficial once a GO-based membrane can be fabricated on a support membrane since the latter usually suffer from fouling, defect problem, permselectivity trade-off, and unstable that renders it unsuitable for industrial application. Thus, it is expected that the resulting GO-based composite membrane can have an exceptional property suitable for industrial application. In this century, the number of research related to graphene oxide membrane significantly increased since it was introduced in 2004, as shown in Figure 1. So, those works necessary to be reviewed to summarise it and as a reflection for the future work. This article will then give a brief summary of the recent studies related to fabrication and performance of GO-based composite membrane for fluid separation.

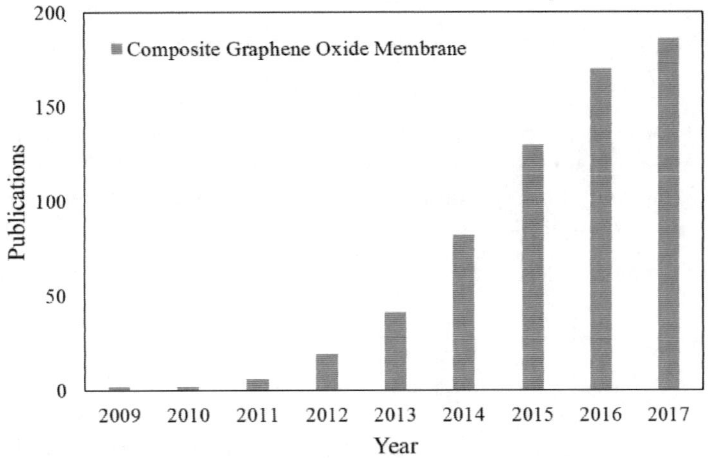

Figure 1. Number of graphene oxide membrane publications indexed by Scopus.

2. Graphene Oxide

Graphene has structure like a honeycomb, which is a two-dimensional material that is built from a hexagonally arrayed sp^2-bonded carbon and only a single-atom in thickness [6]. Meanwhile, graphene oxide (GO) is a chemical derivative of graphene that contains oxygen functionalities such as carboxyl, epoxy, carbonyl and hydroxyl groups [7], as illustrated in Figure 2. The location of these functional groups are both on the edges and basal plane of the GO sheet. The presence of these functional groups renders GO to be attractive to be applied as a membrane because its microstructure, chemical and physical property of the resulting membrane can be positively affected by these functionalities [8]. The strong hydrogen bonds in the graphene oxide nano-sheet hold the sheet compact and form a lamellar structure.

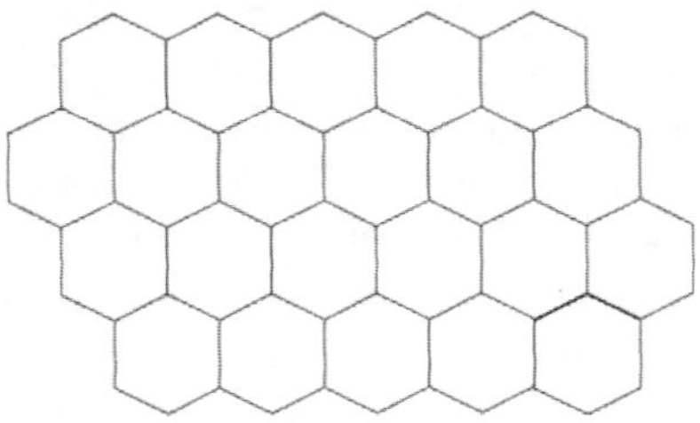

(a)

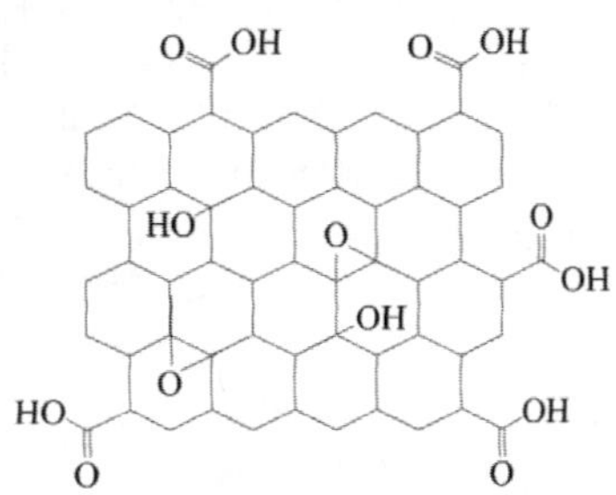

(b)

Figure 2. Chemical structures of (a) graphene and (b) graphene oxide.

The potential of GO to be used as a membrane material was firstly discovered by the work of Nair et al. [9]. In their publication, they revealed that the GO membrane produced on a copper foil -which was etched later on to produce a freestanding GO- was able to let water vapour to pass through while on the other hand, it did not let any gas including helium to pass through. However, helium could start to permeate once the humidity level in the permeation system was increased. They argue that the GO membrane actually consists of pristine graphene capillaries whose pore opening depend on the surrounding humidity. Since then, considerable amount of works have been devoted to GO membrane since its wide-ranging potential applications such as for producing a molecular sieving membrane, water filtration, proton exchange membrane, desalination, etc. [4,10].

3. Preparation and Modification of Composite Graphene Oxide Membrane

3.1. Preparation

Generally, GO can be synthesized through two consecutive steps: oxidation of graphite that is followed by exfoliation of graphite oxide [8,11], as shown in Figure 3. GO was firstly synthesized by Brodie in 1859 [12] using graphite as the starting material which was oxidized using fuming nitric acid with potassium chlorate. Although more practical and convenient, all of these early methods produce toxic gases and not environmentally friendly [11].

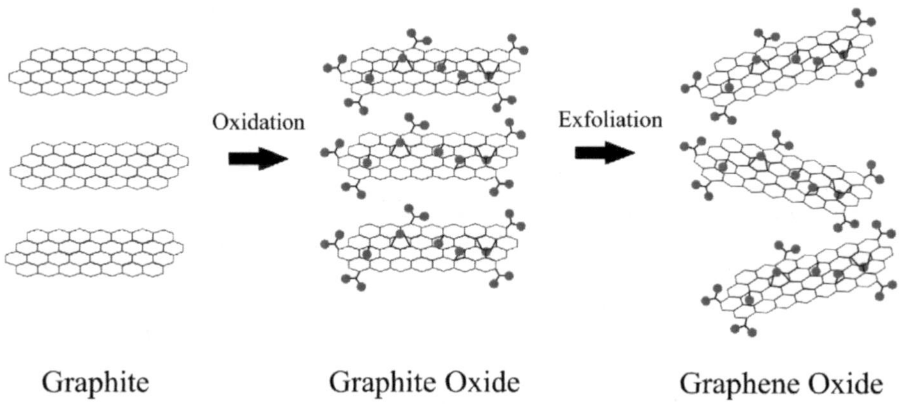

Figure 3. Synthesized process to obtain graphene oxide.

In 20th century, the production process of GO was then significantly improved. In terms of chemical, fuming nitric acid was no longer required in Hofmann's method. This method was then further improved by Hummers and Offeman who used a mixture of potassium permanganate, sulfuric acid and sodium nitrate [14].

Various methods have also been developed recently. For example, Luo et al. [15] have proposed the use of microwave heating method before exfoliation of graphite takes place. The aim of microwave heating is to expand the graphite and reduce its thickness. By doing so, they were able to produce a macroscopic-size GO membrane with the resulting area up to 2000 μm^2. But, the main objective is still how to largely produce a GO so it can be industrially attractive and environmentally friendly [11]. Among of them, the most common method to produce GO currently is through modified Hummer's method. In this process, graphite is firstly oxidized using various oxidizing agent to obtain a monolayer GO sheets. The sheets are then exfoliated to produce GO flakes [5]. Methods to produce graphene oxide membrane composite on a membrane support or substrate are briefly described in this section include filtration, spin coating, surface functionalization, and other techniques.

3.1.1. Filtration. A simple way to prepare a GO membrane is by filtration method. In this technique, GO is homogeneously suspended in a solution. Afterwards, the GO solution filtered using a porous support to obtain a uniform coating. Filtration process can be vacuum- or pressure-assisted. The efficacy of this technique has been proven. For instance, Zhang et al. [16] have prepared a crosslinked

GO membrane on a PVDF support. This technique has also been proven to prepare a GO membrane on inorganic substrate. Li et al. [17] have shown the feasibility to fabricate a GO membrane on an anodized aluminum oxide substrate through this technique. The thickness of the resulted membrane was found to be less than 20 nm, approaching 1.8 nm. The resulting membrane also showed promising separation when tested to separate H_2/CO_2 and H_2/N_2 mixture. Recent developments have also shown the possibility to fabricate a GO membrane on a ceramic hollow fiber by using this technique [18–21]. In this technique, it was found that in order to have a stable GO membrane for real application, the prepared GO membrane needs to be stored in water in order to prevent any shrinkage caused by initial drying [18].

3.1.2. Spin coating. Another simple technique to prepare a GO composite membrane using inorganic substrate is by spin-coating. The benefit of this technique is its ability to produce uniform layer of film compared with other techniques such as drop casting and solvent-induced precipitation. It can also produce a very thin film [22] which is more suitable for membrane application.

This technique has been used to coat GO layer on a SiO2/Si substrate to obtain a very thin layer of GO (about 1 nm in height) [23] and glass and quartz substrate for electronic application [22]. Hu et al. [24] recently prepared a GO membrane on Al_2O_3 substrate using this technique. The resulting composite membrane contains three different layers: membrane support, membrane interlayer, and the active membrane material where the GO lies. Both the support and membrane interlayer parts are fabricated from a mixture of Al_2O_3 powder and polyvinyl alcohol (PVA). In addition, the membrane interlayer also contains dispersant aside from the two main components. The difference between the support and the membrane interlayer is the former is more porous than the later (400 nm compared to 80 nm pore size). As the final step, they spin coated a GO membrane by spin coating followed by drying. The thickness of the resulting membrane was found to be 800 nm and exhibited a good performance regarding its phenol rejection.

3.1.3. Surface functionalization. The essential part of this technique is functionalize the substrate (polymer or inorganic) to make it more active layer so that it can be stronger bonding between GO and the substrate can be established. Unlike the blending method, an active layer is required in this technique to build a covalent bonding between GO layer and the polymer layer. This technique has been used to functionalize a polyamide thin film composite by using 1-ethyl-3-[3-(dimethylamino)propyl] carbondiimide hydrochloride and N-hydroxysuccinimide. This step is necessary to bind GO to the active layer resulting in a functionalized GO-polyamide [25]. Moreover, its process ensures that chemical bonding will established once the modified substrate was dip coated in a GO solution to fabricate a GO membrane. It was also found that the pH of GO solution should be around 5-6 to obtain a robust GO membrane [26]. Another compound that can be used to functionalize alumina substrate is polydopamine [27]. The polydopamine deposited on the alumina substrate founded not to give any additional mass transfer resistance to the alumina disc. After modification, GO layer deposited on the substrate through vacuum filtration.

3.1.4. Other possible techniques. A composite GO membrane can also be prepared by direct blending with another material, usually polymer. After blending process takes place, the resulting dope solution is usually cast in a porous support. This will result a mixed matrix composite membrane with dispersed GO particles. This method has been used to prepare a GO-PBI membrane for PEM fuel cell membrane [28], GO-PVDF for membrane bioreactor [29], and GO-PVA [30]. This technique can be slightly modified by employing porous inorganic support when casting the resulting dope solution. The casting process itself can be completed using drop-casting or spin coating technique that have been established for other GO-composite membrane.

The mechanical strength of composite graphene oxide membrane has discovered by the work of some researchers. [31–36]. The improvements of tensile strength and Young's modulus are attributed to the homogenous dispersion of GO nano-sheet at the surface or matrix membrane. Furthermore, there is a strong interaction between GO oxygen-containing functional groups and support or substrate membrane via hydrogen bonding [35].

3.2. Modification

Graphene oxide-based membrane can be constructed in the form of freestanding or self-assembly GO layer [37,38], modified support membrane [39–41], thin film composite [25,42–45], and mixed matrix membrane array [46–49]. Freestanding GO membrane utilises GO as a separating layer directly. Self-assembly or freestanding GO has shown ultrafast permeance and high rejection of solute [50]. In another hand, this multilayer GO had a weak mechanical strength if applied in high-pressure operation. Furthermore, the swelling effect in freestanding GO due to intercalation of water molecules in the interlayer spacing leads to unstable framework due to increase of interlayer spacing *d*. [51–53].

Improve mechanical strength and stability of membrane could be approached by chemical crosslinking of GO layer [54]. Chemicals such polydopamine [55], isocyanate [56,57], metal ions [58], diamine monomers [59], glutaraldehyde [60] have been reported to elevate mechanical resistance and stability of the membrane. Covalent bonding between GO layer and crosslinker agent sufficiently suppress the interlayer spacing *d* [59]. Hence, the stability and integrity of GO membrane could be maintained.

Incorporated-graphene oxide into material support has been reported enhance the separation performance and property of membrane. The amount of graphene oxide which embeds in support membrane will affect permeance, solute rejection, mechanical strength, and pH resistive of membrane. The higher amount of GO added into membrane could enhance the mechanical integrity of membrane [32], however it will induce aggregation of GO which blocking the water/gas permeation. [61,62]. The lower GO addition into membrane the more sensitive the as-prepared GO against pH [32]. So, the proper quantity of GO to be embedded into support membrane need to be considered to achieve satisfied separation performance and stability of membrane.

4. Application and Separation Performance of Composite Inorganic Graphene Oxide Membrane

4.1. Gas Separation

Graphene oxide membranes have been shown to be capable of performing gas separation process. This phenomenon has been discovered and studied by numerous scientists through either experimental and simulation. Graphene oxide was incorporated into polymer support layer with expectancy to alleviate the trade-off between permeability-selectivity and plasticization [63,64]. Thus, graphene oxide was an appropriate nanomaterial to be used as fillers in the gas separation membrane [65].

The gases permselectivity are relying on the diffusivity and solubility of the gas in a membrane. Several gases have been identified their permeation performance through graphene oxide-based membranes, which are the gas separation performance follows in order $CO_2 > H_2 > CH_4 > N_2$ [66–68]. In the availability of water, the permeability of carbon dioxide gas significantly increases compared in dry condition. The high CO_2 permeance in GO membrane can be assumed in the high CO_2 diffusivity, and the high CO_2 selectivity can be ascribed to the high CO_2 solubility [66,69]. This property of GO membrane corresponds to carbon dioxide gas give an amplified that GO membrane can be utilised for the CO_2 capture [70–74].

Selective gas diffusion can be achieved via controlling interlayer spacing between GO layer and nano-pores between the inter-edge of GO sheets [75,76]. Romanos et al. have successfully carried out variation filtration rate of the as-prepared GO suspension. The finding was that fast filtration rate led to a random rearrangement of GO layer, on the other hand very slow filtration resulted in order structures. The as-prepared GO membranes by slow filtration were more hydrophobic than another GO which prepared by fast filtration. The more hydrophobic membrane result in higher water vapour permeation. So, this finding offers the opportunities of GO membrane to be applied for distillation process [77].

The support membrane in the performance of gas separation membrane also possesses a key role. The support membranes have to have thermal stability and swelling resistant. Incorporated graphene oxide into support polymer membrane has obviously enhanced the performance of membrane for gas separation. The permeability-selectivity, include thermal stability of membrane was significantly improve [78,79]. Polymer-support membrane which intercalated graphene oxide have been successfully produced by some researchers include polysulfone [80–82], Pebax [71,83], polyether sulfone [84], and polyimide [85]. Separation performances of several GO membranes in gas separation are summarized in Table 1.

Table 1. Separation performances of several GO membranes in gas separation.

No.	GO-based membrane	Permeance (mol/m^2sPa)	Selectivity	Ref.
1.	GO-AAO	$P_{H2} = 3.400 \cdot 10^{-7}$	$\alpha_{(H2/CO2)} = 240$	[86]
2.	Annealed-GO-polycarbonate	$P_{H2} = 1.758 \cdot 10^{-7}$	$\alpha_{(H2/CO2)} = 10$	[87]
3.	Aminated-GO/polyethersulfone mixed matrix membrane	$P_{CO2} = 4.421 \cdot 10^{-9}$	$\alpha_{(CO2/CH4)} = 37.88$	[84]
4.	GO-UiO66-NH_2-S	$P_{H2} = 3.900 \cdot 10^{-8}$	$\alpha_{(H2/CO2)} = 6.35$	[88]
5.	GO/α-Al_2O_3 ceramic hollow fiber membrane	$P_{H2} = 1.300 \cdot 10^{-7}$	$\alpha_{(H2/CO2)} = 15$	[89]
6.	ODA-GO/polydimethylsiloxane	$P_{C3H8} = 6.355 \cdot 10^{-7}$	$\alpha_{(C3H8/N2)} = 67$	[90]

4.2. Water Treatment

Graphene oxide is known can be well-dispersed in water because of the electrostatic repulsion between the ionized functional group in GO nano-sheet [63]. Wei et al. have examined the fast permeation of water through graphene oxide, water molecules pass through the nano-porous between the edge of GO sheets and between the interlayer spacing of GO layer [91]. Because of its characteristic and behaviour in water, the application of graphene oxide-based membrane in water purification including heavy metal removal [47,92], oil/water separation [38,93], dye removal [94] and desalination [95,96] is highly promising.

Embedded-graphene oxide in membrane support has proven to enhance the water permeation flux and solute rejection. Moreover, the fouling resistance and antibacterial properties of membrane become more favourable than the bare membrane [97–104]. Hydrophilicity properties of GO membrane leading to high water flux. The enhancement of hydrophilicity is induced as a result of the surface functionalization of graphene oxide membrane [105]. The high solute rejection was explained by three mechanisms of filtration, include size exclusion [106,107], electrostatic interaction [106,108,109], and/or adsorption [110,111]. Separation performances of several GO membranes in water treatment are summarized in Table 2.

Table 2. Separation performances of several GO membranes in water purification.

No.	GO-based membrane	Application	Testing condition	Water flux (L/m^2hbar)	Rejection (%)	Ref.
1.	COOH-GO self assembly membrane (NF)	Desalination	$NaCl_{feed} = 2000$ ppm, P = 15 bar	4.89	39.16	[96]
2.	rGO/TiO2 polysulfone TFC membrane (RO)	Desalination	$NaCl_{feed} = 2000$ ppm, P = 15 bar	3.42	99.45	[61]
3.	Aminated-GO/polyamide TFC membrane (RO)	Desalination	$NaCl_{feed} = 2000$ ppm, P = 15.5 bar	0.903	96.4	[112]
4.	GO-palygorskite nanohybrid membrane	Oil/Water Separation	Stirring rate = 200 rpm, P = 0.5 bar	3734	99.9	[38]
5.	GO/Torlon hollow fiber membrane	Heavy Metal Removsl	$Pb^{2+}_{feed} = 1000$ ppm, P = 3 bar	4.7	95.88	[113]
6.	GO-IPDI membrane	Dye Removal	Initial feed (congo red) = 20 mL, pH 7, P = 1 bar	80~100	99.3	[57]

4.3. Electrochemical Energy

Nowadays, clean energy production was preferably carried out due to efficient and environmentally friendly process. Electrochemical energy sources divided into several types, such as fuel cell. Membrane technology in these processes has based on electro-ionic separation, such as cation exchange membrane, anion exchange membrane, or a bipolar membrane. Proton exchange membrane in fuel cell application has to give a good performance such in proton conductivity, mechanical and thermal stability, oxidation stability, water uptake and low ion permeability.

The addition of nano-additives or nano-fillers (e.g. graphene oxide) into polymer membrane is one of the alternatives to improve the performance (ion exchange capacity, water uptake, and proton conductivity) of proton exchange membrane [114–118]. Jiang et al. found that the introduction of Si-GOs into polymer membrane improve the mechanical stability without diminished the proton conductivity [119]. The mechanical stability of polymer membrane with the inclusion of graphene oxide also has been demonstrated by various authors [115,120–123]. Separation performances of several GO membranes in fuel cells application are summarized in Table 3.

Table 3. Separation performances of several GO membranes in fuel cells application.

No.	GO-based membrane	Application	Testing condition	Proton conductivity (S cm^{-1})	Power density (mW cm^{-2})	Ref.
1.	Sulfonated-GO/Nafion membrane	Direct Methanol Fuel Cells	Methanol = 1 M	0.0363	42.9	[124]
2.	Freestanding GO membrane	Polymer Electrolyte Membrane Fuel Cells	T = 70^{0}C RH = 100%	0.0011	33.8	[125]
3.	Aryl sulfonated-GO/poly(vinyl alcohol) mixed matrix membrane	Polymer Electrolyte Membrane Fuel Cells	T = 30^{0}C	0.0500	16.15	[126]
4.	Sulfonic acid-GO/sulfonated polyether ether ketone composite membrane	Polymer Electrolyte Membrane Fuel Cells	T = 80^{0}C RH = 30%	0.0550	378	[127]
5.	GO/sulfonated polyether ether ketone-histidine composite membrane	Direct Methanol Fuel Cells	T = 25^{0}C RH = 100% Methanol = 2 M	0.0694	43	[128]

5. Summary and Outlook

Composite Inorganic Graphene Oxide Membrane with all its exceptional physicochemical properties has gained increased interest recently. Graphene oxide is hydrophilic, its property due to oxygen-containing functional groups in the GO sheet. When fabricated as a membrane, GO can act as a selective layer that selectively let some components pass through while rejecting the rest.

Graphene oxide membrane can be constructed in the form of self-assembly or free-standing GO layer, modified support membrane, thin film composite or mixed matrix membrane. Composite inorganic graphene oxide membrane can be prepared by filtration vacuum, spin/spray/dip coating, surface functionalization and other possible technique such as blending GO into a polymer matrix.

According to several reported studies, composite graphene oxide membranes have showed the enhanced separation performance in various applications such as gas separation, water treatment, and fuel cell. However, the stability of composite GO membrane needs further development. Among others approached by adding crosslinker to bind the GO inter-sheets. Another aspect need to be considered is the quantity of GO added into support membrane. With expectancy to enhance the stability and separation performance of membrane, an appropriate quantity of GO added into support membrane is important to determine.

6. References

[1] Wenten I G 2003 Recent development in membrane science and its industrial applications *Songklanakarin J. Sci. Technol.* **24** 1009–23
[2] Le N L and Nunes S P 2016 Materials and membrane technologies for water and energy sustainability *Sustain. Mater. Technol.* **7** 1–28
[3] Zhang Y and Chung T 2017 Graphene oxide membranes for nanofiltration *Curr. Opin. Chem. Eng.* **16** 9–15
[4] Joshi R K, Alwarappan S, Yoshimura M, Sahajwalla V and Nishina Y 2015 Graphene oxide: The new membrane material *Appl. Mater. Today* **1** 1–12
[5] Chong J, Wang B and Li K 2016 Graphene oxide membranes in fluid separations *Curr. Opin. Chem. Eng.* **12** 98–105
[6] Kim H, Abdala A A and MacOsko C W 2010 Graphene/polymer nanocomposites *Macromolecules* **43** 6515–30
[7] Dreyer D R, Park S, Bielawski C W and Ruoff R S 2010 The chemistry of graphene oxide *Chem. Soc. Rev.* **39** 228–40
[8] Liu G, Jin W and Xu N 2015 Graphene-based membranes *Chem. Soc. Rev.* **44** 5016–30
[9] R. R. Nair, H. A. Wu, P. N. Jayaram, I. V. Grigorieva and A. K. Geim 2012 *Unimpeded Permeation of Water Through Helium-Leak–Tight Graphene-Based Membranes* vol 335(American Association for the Advancement of Science)
[10] Abraham J, Vasu K S, Williams C D, Gopinadhan K, Su Y, Cherian C T, Dix J, Prestat E, Haigh S J, Grigorieva I V., Carbone P, Geim A K and Nair R R 2017 Tunable sieving of ions using graphene oxide membranes *Nat. Nanotechnol.* **12** 546–50
[11] Ma J, Ping D and Dong X 2017 Recent Developments of Graphene Oxide-Based Membranes: A Review *Membranes (Basel).* **7** 52
[12] Brodie B C 1859 On the Atomic Weight of Graphite *Philos. Trans. R. Soc. London* **149** 249–59
[13] Staudenmaier L 1898 Verfahren zur Darstellung der Graphitsäure *Eur. J. Inorg. Chem.* **31** 1481–7
[14] Hummers W S and Offeman R E 1958 Preparation of Graphitic Oxide *J. Am. Chem. Soc.* **80** 1339–1339
[15] Luo Z, Lu Y, Somers L A and Johnson A T . 2009 High Yield Preparation of Macroscopic Graphene Oxide Membranes High Yield Preparation of Macroscopic Graphene Oxide Membranes *J. Am. Chem. Soc.* **131** 898–9
[16] Zhang P, Gong J, Zeng G, Deng C, Yang H, Liu H and Huan S 2017 Cross-Linking to Prepare Composite Graphene Oxide-Framework Membranes with High-Flux for Dyes and Heavy Metal Ions Removal *Chem. Eng. J.* **322** 657-666
[17] Li H, Song Z, Zhang X, Huang Y, Li S, Mao Y, Ploehn H J, Bao Y and Yu M 2013 Ultrathin, molecular-sieving graphene oxide membranes for selective hydrogen separation. *Science.* **342** 95–8
[18] Aba N F D, Chong J Y, Wang B, Mattevi C and Li K 2015 Graphene oxide membranes on ceramic hollow fibers - Microstructural stability and nanofiltration performance *J. Memb. Sci.* **484** 87–94
[19] Huang K, Liu G, Lou Y, Dong Z, Shen J and Jin W 2014 A Graphene Oxide Membrane with Highly Selective Molecular Separation of Aqueous Organic Solution *Angew. Chemie Int. Ed.* **53** 6929–32
[20] Zhu J, Meng X, Zhao J, Jin Y, Yang N and Zhang S 2017 Facile hydrogen/nitrogen separation through graphene oxide membranes supported on YSZ ceramic hollow fibers *J. Memb. Sci.* **535** 143–50
[21] Huang K, Yuan J, Shen G, Liu G and Jin W 2017 Graphene oxide membranes supported on the ceramic hollow fibre for efficient H2 recovery *Chinese J. Chem. Eng.* **25** 752–9
[22] Becerril H A, Mao J, Liu Z, Stoltenberg R M, Bao Z and Chen Y 2008 Evaluation of Solution-

Processed Reduced Graphene Oxide Films as Transparent Conductors *ACS Nano* **2** 463–70
[23] Luo Z, Lu Y, Somers L A and Johnson A T C 2009 High Yield Preparation of Macroscopic Graphene Oxide Membranes *J. Am. Chem. Soc.* **131** 898–9
[24] Hu X, Yu Y, Ren S, Lin N, Wang Y and Zhou J 2017 Highly efficient removal of phenol from aqueous solutions using graphene oxide/Al2O3 composite membrane *J. Porous Mater.* 1–8
[25] Perreault F, Tousley M E and Elimelech M 2013 Thin-Film Composite Polyamide Membranes Functionalized with Biocidal Graphene Oxide Nanosheets *Environ. Sci. Technol. Lett.* **1** 71–6
[26] Lou Y, Liu G, Liu S, Shen J and Jin W 2014 A facile way to prepare ceramic-supported graphene oxide composite membrane via silane-graft modification *Appl. Surf. Sci.* **307** 631–7
[27] Xu K, Feng B, Zhou C and Huang A 2016 Synthesis of highly stable graphene oxide membranes on polydopamine functionalized supports for seawater desalination *Chem. Eng. Sci.* **146** 159–65
[28] Xue C, Zou J, Sun Z, Wang F, Han K and Zhu H 2014 Graphite oxide/functionalized graphene oxide and polybenzimidazole composite membranes for high temperature proton exchange membrane fuel cells *Int. J. Hydrogen Energy* **39** 7931–9
[29] Zhao C, Xu X, Chen J, Wang G and Yang F 2014 Highly effective antifouling performance of PVDF/graphene oxide composite membrane in membrane bioreactor (MBR) system *Desalination* **340** 59–66
[30] Lecaros R L G, Mendoza G E J, Hung W S, An Q F, Caparanga A R, Tsai H A, Hu C C, Lee K R and Lai J Y 2017 Tunable interlayer spacing of composite graphene oxide-framework membrane for acetic acid dehydration *Carbon N. Y.* **123** 660–7
[31] Sridi N, Lebental B, Azevedo J, Gabriel J C P and Ghis A 2013 Electrostatic method to estimate the mechanical properties of suspended membranes applied to nickel-coated graphene oxide *Appl. Phys. Lett.* **103**
[32] Jiang Z, Xia D, Li Y, Li J, Li Q, Chen M, Huang Y, Besenbacher F and Dong M 2013 Facilitating the mechanical properties of a high-performance pH-sensitive membrane by cross-linking graphene oxide and polyacrylic acid. *Nanotechnology* **24** 335704
[33] Lee S, Choi B G, Choi D and Park H S 2014 Nanoindentation of annealed Nafion/sulfonated graphene oxide nanocomposite membranes for the measurement of mechanical properties *J. Memb. Sci.* **451** 40–5
[34] Thampi S, Muthuvijayan V and Parameswaran R 2015 Mechanical characterization of high-performance graphene oxide incorporated aligned fibroporous poly(carbonate urethane) membrane for potential biomedical applications *J. Appl. Polym. Sci.* **132** 41808–16
[35] Gan S, Zakaria S, Chia C H, Chen R S and Jeyalaldeen N 2015 Physico-mechanical properties of a microwave-irradiated kenaf carbamate/graphene oxide membrane *Cellulose* **22** 3851–63
[36] Wang X, Wang X, Xiao P, Li J, Tian E, Zhao Y and Ren Y 2016 High water permeable free-standing cellulose triacetate/graphene oxide membrane with enhanced antibiofouling and mechanical properties for forward osmosis *Colloids Surfaces A Physicochem. Eng. Asp.* **508** 327–35
[37] Xu W L, Fang C, Zhou F, Song Z, Liu Q, Qiao R and Yu M 2017 Self-Assembly: A Facile Way of Forming Ultrathin, High-Performance Graphene Oxide Membranes for Water Purification *Nano Lett.* **17** 2928–33
[38] Zhao X, Su Y, Liu Y, Li Y and Jiang Z 2016 Free-Standing Graphene Oxide-Palygorskite Nanohybrid Membrane for Oil/Water Separation *ACS Appl. Mater. Interfaces* **8** 8247–56
[39] Gao Y, Hu M and Mi B 2014 Membrane surface modification with TiO2-graphene oxide for enhanced photocatalytic performance *J. Memb. Sci.* **455** 349–56
[40] Huang X, Marsh K L, McVerry B T, Hoek E M V and Kaner R B 2016 Low-Fouling Antibacterial Reverse Osmosis Membranes via Surface Grafting of Graphene Oxide *ACS Appl. Mater. Interfaces* **8** 14334–8
[41] Lou Y, Liu G, Liu S, Shen J and Jin W 2014 Applied Surface Science A facile way to prepare ceramic-supported graphene oxide composite membrane via silane-graft modification *Appl.*

Surf. Sci. **307** 631–7

[42] Yin J, Zhu G and Deng B 2016 Graphene oxide (GO) enhanced polyamide (PA) thin-film nanocomposite (TFN) membrane for water purification *Desalination* **379** 93–101

[43] Shen L, Xiong S and Wang Y 2016 Graphene oxide incorporated thin-film composite membranes for forward osmosis applications *Chem. Eng. Sci.* **143** 194–205

[44] Chae H-R, Lee C-H, Park P-K, Kim I-C and Kim J-H 2016 Synergetic effect of graphene oxide nanosheets embedded in the active and support layers on the performance of thin-film composite membranes *J. Memb. Sci.* **525** 99–106

[45] Park M J, Phuntsho S, He T, Nisola G M, Tijing L D, Li X M, Chen G, Chung W J and Shon H K 2015 Graphene oxide incorporated polysulfone substrate for the fabrication of flat-sheet thin-film composite forward osmosis membranes *J. Memb. Sci.* **493** 496–507

[46] Kaleekkal N J, Thanigaivelan A, Rana D and Mohan D 2016 Studies on carboxylated graphene oxide incorporated polyetherimide mixed matrix ultrafiltration membranes *Mater. Chem. Phys* **186** pp. 146-158

[47] Mukherjee R, Bhunia P and De S 2016 Impact of graphene oxide on removal of heavy metals using mixed matrix membrane *Chem. Eng. J.* **292** 284–97

[48] Mohd Akhair S S, Harun Z, Jamalludin M R, Shuhor M F, Kamarudin N H, Yunos M Z, Ahmad A and Azhar M F H 2017 Polymer Mixed Matrix Membrane with Graphene Oxide for Humic Acid Performances *Chemical Engineering Transactions* **56** 697–702

[49] Zhao L, Cheng C, Chen Y F, Wang T, Du C H and Wu L G 2015 Enhancement on the permeation performance of polyimide mixed matrix membranes by incorporation of graphene oxide with different oxidation degrees *Polym. Adv. Technol.* **26** 330–7

[50] Jang J H and Lee J 2015 Ultrathin graphene oxide membranes for water purification *2015 IEEE 15th International Conference on Nanotechnology (IEEE-NANO)* pp 212–5

[51] Zheng S, Tu Q, Urban J J, Li S and Mi B 2017 Swelling of Graphene Oxide Membranes in Aqueous Solution: Characterization of Interlayer Spacing and Insight into Water Transport Mechanisms *ACS Nano* **11** 6440–50

[52] Xu X L, Lin F W, Du Y, Zhang X, Wu J and Xu Z K 2016 Graphene Oxide Nanofiltration Membranes Stabilized by Cationic Porphyrin for High Salt Rejection *ACS Appl. Mater. Interfaces* **8** 12588–93

[53] Heo Y, Im H and Kim J 2013 The effect of sulfonated graphene oxide on Sulfonated Poly (Ether Ether Ketone) membrane for direct methanol fuel cells *J. Memb. Sci.* **425–426** 11–22

[54] Jia Z and Shi W 2016 Tailoring permeation channels of graphene oxide membranes for precise ion separation *Carbon N. Y.* **101** 290–5

[55] He Y, Wang J, Zhang H, Zhang T, Zhang B, Cao S and Liu J 2014 Polydopamine-modified graphene oxide nanocomposite membrane for proton exchange membrane fuel cell under anhydrous conditions *J. Mater. Chem. A* **2** 9548

[56] Zhao H, Wu L, Zhou Z, Zhang L and Chen H 2013 Improving the antifouling property of polysulfone ultrafiltration membrane by incorporation of isocyanate-treated graphene oxide *Phys. Chem. Chem. Phys.* **15** 9084

[57] Zhang P, Gong J, Zeng G, Deng C, Yang H and Liu H 2017 Cross-linking to prepare composite graphene oxide-framework membranes with high-flux for dyes and heavy metal ions removal *Chem. Eng. J.* **322** 657–66

[58] Rao Z, Feng K, Tang B and Wu P 2017 Surface decoration of amino-functionalized metal-organic framework/graphene oxide composite onto polydopamine-coated membrane substrate for highly efficient heavy metal removal *ACS Appl. Mater. Interfaces* **9** 2594–605

[59] Hung W, Tsou C, Guzman M De, An Q, Liu Y, Zhang Y, Hu C, Lee K and Lai J 2014 Cross-Linking with Diamine Monomers To Prepare Composite Graphene Oxide-Framework Membranes with Varying d-Spacing *Chem. Mater.* **26** 2983–90

[60] Sarkar G, Saha N R, Roy I, Bhattacharyya A, Adhikari A, Rana D, Bhowmik M, Bose M, Mishra R and Chattopadhyay D 2016 Cross-linked methyl cellulose/graphene oxide rate controlling membranes for in vitro and ex vivo permeation studies of diltiazem hydrochloride *RSC Adv.* **6** 36136–45

[61] Safarpour M, Khataee A and Vatanpour V 2015 Effect of reduced graphene oxide/TiO2 nanocomposite with different molar ratios on the performance of PVDF ultrafiltration membranes *Sep. Purif. Technol.* **140** 32–42
[62] Wu X, Field R W, Wu J J and Zhang K 2017 Polyvinylpyrrolidone modified graphene oxide as a modifier for thin film composite forward osmosis membranes *J. Memb. Sci.* **540** 251–60
[63] Yoo B M, Shin J E, Lee H D and Park H B 2017 Graphene and graphene oxide membranes for gas separation applications *Curr. Opin. Chem. Eng.* **16** 39–47
[64] Zahri K, Goh P S and Ismail A F 2016 The incorporation of graphene oxide into polysulfone mixed matrix membrane for CO2/CH4 separation *IOP Conf. Ser. Earth Environ. Sci.* **36** 12007
[65] Li X, Cheng Y, Zhang H, Wang S, Jiang Z, Guo R and Wu H 2015 Efficient CO2 capture by functionalized graphene oxide nanosheets as fillers to fabricate multi-permselective mixed matrix membranes *ACS Appl. Mater. Interfaces* **7** 5528–37
[66] Kim H W, Yoon H W, Yoo B M, Park J S, Gleason K L, Freeman B D and Park H B 2014 High-performance CO2-philic graphene oxide membranes under wet-conditions *Chem. Commun.* **50** 13563–6
[67] Karunakaran M, Shevate R, Kumar M and Peinemann K 2015 CO2-selective PEO–PBT (PolyActivet)/graphene oxide composite membranes *Chem. Commun.* **51** 14187–90
[68] Shen J, Liu G, Huang K, Jin W, Lee K R and Xu N 2015 Membranes with fast and selective gas-transport channels of laminar graphene oxide for efficient CO2 capture *Angew. Chemie - Int. Ed.* **54** 578–82
[69] Koolivand H, Sharif A, Kashani M R, Karimi M, Salooki M K and Semsarzadeh M A 2014 Functionalized graphene oxide/polyimide nanocomposites as highly CO2-selective membranes *J. Polym. Res.* **21** 599
[70] Karunakaran M, Villalobos L F, Kumar M, Shevate R, Akhtar F H, Peinemann K-V, Noble R D, Bauer E M, Nasillo G, Caponetti E, Choi J Y and Park H B 2017 Graphene oxide doped ionic liquid ultrathin composite membranes for efficient CO_2 capture *J. Mater. Chem. A* **5** 649–56
[71] Dai Y, Ruan X, Yan Z, Yang K, Yu M, Li H, Zhao W and He G 2016 Imidazole functionalized graphene oxide/PEBAX mixed matrix membranes for efficient CO2 capture *Sep. Purif. Technol.* **166** 171–80
[72] Shen Y, Wang H, Liu J and Zhang Y 2015 Enhanced Performance of a Novel Polyvinyl Amine/Chitosan/Graphene Oxide Mixed Matrix Membrane for CO2 Capture *ACS Sustain. Chem. Eng.* **3** 1819–29
[73] Lim M, Choi Y, Kim J, Kim K, Shin H, Kim J, Myung D and Lee J 2017 Cross-linked graphene oxide membrane having high ion selectivity and antibacterial activity prepared using tannic acid-functionalized graphene oxide and polyethyleneimine *J. Memb. Sci.* **521** 1–9
[74] Dong G, Zhang Y, Hou J, Shen J and Chen V 2016 Graphene Oxide Nanosheets Based Novel Facilitated Transport Membranes for Efficient CO 2 Capture *Ind. Eng. Chem. Res.* **55** 5403–14
[75] Kim H W, Yoon H W, Yoon S-M, Yoo B M, Ahn B K, Cho Y H, Shin H J, Yang H, Paik U, Kwon S, Choi J-Y and Park H B 2013 Selective Gas Transport Through Few-Layered Graphene and Graphene Oxide Membranes *Science (80-.).* **342** 91–5
[76] Jiao S and Xu Z 2015 Selective gas diffusion in graphene oxides membranes: A molecular dynamics simulations study *ACS Appl. Mater. Interfaces* **7** 9052–9
[77] Romanos G, Pastrana-Martinez L M, Tsoufis T, Athanasekou C, Galata E, Katsaros F, Favvas E, Beltsios K G, Siranidi E, Falaras P, Psycharis V and Silva A M T 2015 A facile approach for the development of fine-tuned self-standing graphene oxide membranes and their gas and vapor separation performance *J. Memb. Sci.* **493** 734–47
[78] Wang T, Zhao L, Shen J N, Wu L G and Van Der Bruggen B 2015 Enhanced Performance of Polyurethane Hybrid Membranes for CO2 Separation by Incorporating Graphene Oxide: The Relationship between Membrane Performance and Morphology of Graphene Oxide *Environ. Sci. Technol.* **49** 8004–11

[79] Ha H, Park J, Ando S, Kim C Bin, Nagai K, Freeman B D and Ellison C J 2016 Gas permeation and selectivity of poly(dimethylsiloxane)/graphene oxide composite elastomer membranes *J. Memb. Sci.* **518** 131–40

[80] Sarfraz M and Ba-Shammakh M 2015 Synergistic effect of incorporating ZIF-302 and graphene oxide to polysulfone to develop highly selective mixed-matrix membranes for carbon dioxide separation from wet post-combustion flue gases *J. Ind. Eng. Chem. **36** 154–162*

[81] Sarfraz M and Ba-Shammakh M 2016 Synergistic effect of adding graphene oxide and ZIF-301 to polysulfone to develop high performance mixed matrix membranes for selective carbon dioxide separation from post combustion flue gas *J. Memb. Sci.* **514** 35–43

[82] Zahri K, Wong K C, Goh P S and Ismail A F 2016 Graphene oxide/polysulfone hollow fiber mixed matrix membranes for gas separation *RSC Adv.* **6** 89130–9

[83] Dong G, Hou J, Wang J, Zhang Y, Chen V and Liu J 2016 Enhanced CO2/N2 separation by porous reduced graphene oxide/Pebax mixed matrix membranes *J. Memb. Sci.* **520** 860–8

[84] Ebrahimi S, Mollaiy-Berneti S, Asadi H, Peydayesh M, Akhlaghian F and Mohammadi T 2016 PVA/PES-amine-functional graphene oxide mixed matrix membranes for CO2/CH4 separation: Experimental and modeling *Chem. Eng. Res. Des.* **109** 647–56

[85] Wang T, Cheng C, Wu L G, Shen J N, Van Der Bruggen B, Chen Q, Chen D and Dong C Y 2017 Fabrication of Polyimide Membrane Incorporated with Functional Graphene Oxide for CO2 Separation: The Effects of GO Surface Modification on Membrane Performance *Environ. Sci. Technol.* **51** 6202–10

[86] Chi C, Wang X, Peng Y, Qian Y, Hu Z, Dong J and Zhao D 2016 Facile Preparation of Graphene Oxide Membranes for Gas Separation *Chem. Mater.* **28** 2921–7

[87] Achari A and Eswaramoorthy M 2016 Casting molecular channels through domain formation: high performance graphene oxide membranes for H2/CO2 separation *J. Mater. Chem. A* **4** 7560–4

[88] Jia M, Feng Y, Liu S, Qiu J and Yao J 2017 Graphene oxide gas separation membranes intercalated by UiO-66-NH2 with enhanced hydrogen separation performance *J. Memb. Sci.* **539** 172–7

[89] Huang K, Yuan J, Shen G, Liu G and Jin W 2017 Graphene oxide membranes supported on the ceramic hollow fibre for efficient H2 recovery *Chinese J. Chem. Eng.* **25** 752–9

[90] Shen G, Zhao J, Guan K, Shen J and Jin W 2017 Highly Efficient Recovery of Propane by Mixed-Matrix Membrane via Embedding Functionalized GO Nanosheets into PDMS *Am. Inst. Chem. Eng. J.* **63** 3501–10

[91] Wei N, Peng X and Xu Z 2014 Understanding water permeation in graphene oxide membranes *ACS Appl. Mater. Interfaces* **6** 5877–83

[92] Zhang Y, Zhang S, Gao J and Chung T S 2016 Layer-by-layer construction of graphene oxide (GO) framework composite membranes for highly efficient heavy metal removal *J. Memb. Sci.* **515** 230–7

[93] Hu X, Yu Y, Zhou J, Wang Y, Liang J, Zhang X, Chang Q and Song L 2015 The improved oil/water separation performance of graphene oxide modified Al2O3 microfiltration membrane *J. Memb. Sci.* **476** 200–4

[94] Zhu Z, Wang L, Xu Y, Li Q, Jiang J and Wang X 2017 Preparation and characteristics of graphene oxide-blending PVDF nanohybrid membranes and their applications for hazardous dye adsorption and rejection *J. Colloid Interface Sci.* **504** 429–39

[95] Ali M E A, Wang L, Wang X and Feng X 2016 Thin film composite membranes embedded with graphene oxide for water desalination *Desalination* **386** 67–76

[96] Yuan Y, Gao X, Wei Y, Wang X, Wang J, Zhang Y and Gao C 2017 Enhanced desalination performance of carboxyl functionalized graphene oxide nanofiltration membranes *Desalination* **405** 29–39

[97] Zhang J, Xu Z, Shan M, Zhou B, Li Y, Li B, Niu J and Qian X 2013 Synergetic effects of oxidized carbon nanotubes and graphene oxide on fouling control and anti-fouling mechanism of polyvinylidene fluoride ultrafiltration membranes *J. Memb. Sci.* **448** 81–92

[98] Wang J, Zhao C, Wang T, Wu Z, Li X and Li J 2016 Graphene oxide polypiperazine-amide

nanofiltration membrane for improving flux and anti-fouling in water purification *RSC Adv.* **6** 82174–85

[99] Salehi H, Rastgar M and Shakeri A 2017 Anti-Fouling and High Water Permeable Forward Osmosis Membrane Fabricated via Layer by Layer Assembly of Chitosan/Graphene Oxide *Appl. Surf. Sci.* **413** 99–108

[100] Kumar M, McGlade D, Ulbricht M and Lawler J 2015 Quaternized polysulfone and graphene oxide nanosheet derived low fouling novel positively charged hybrid ultrafiltration membranes for protein separation *RSC Adv.* **5** 51208–19

[101] Safarpour M, Vatanpour V, Khataee A and Esmaeili M 2015 Development of a novel high flux and fouling-resistant thin film composite nanofiltration membrane by embedding reduced graphene oxide/TiO2 *Sep. Purif. Technol.* **154** 96–107

[102] Vatanpour V, Shockravi A, Zarrabi H, Nikjavan Z and Javadi A 2015 Fabrication and characterization of anti-fouling and anti-bacterial Ag-loaded graphene oxide/polyethersulfone mixed matrix membrane *J. Ind. Eng. Chem.* **30** 342–52

[103] Hegab H M, Wimalasiri Y, Ginic-Markovic M and Zou L 2015 Improving the fouling resistance of brackish water membranes via surface modification with graphene oxide functionalized chitosan *Desalination* **365** 99–107

[104] Zambare R S, Dhopte K B, Patwardhan A V. and Nemade P R 2017 Polyamine functionalized graphene oxide polysulfone mixed matrix membranes with improved hydrophilicity and anti-fouling properties *Desalination* **403** 24–35

[105] Ayyaru S and Ahn Y 2017 Application of sulfonic acid group functionalized graphene oxide to improve hydrophilicity , permeability , and antifouling of PVDF nanocomposite ultra filtration membranes *J. Memb. Sci.* **525** 210–9

[106] Hu M and Mi B 2013 Enabling Graphene Oxide Nanosheets as Water Separation Membranes *Environ. Sci. Technol.* **47** 3715–23

[107] Fang Q, Zhou X, Deng W, Zheng Z and Liu Z 2016 Freestanding bacterial cellulose- graphene oxide composite membranes with high mechanical strength for selective ion permeation *Sci. Rep.* **6** 1–11

[108] Liu G, Ye H, Li A, Zhu C, Jiang H and Liu Y 2016 Graphene oxide for high-efficiency separation membranes : Role of electrostatic interactions *Carbon N. Y.* **110** 56–61

[109] Fathizadeh M, Xu W L, Zhou F, Yoon Y and Yu M 2017 Graphene Oxide: A Novel 2-Dimensional Material in Membrane Separation for Water Purification *Adv. Mater. Interfaces* **4**(5) 1600918

[110] Tan P, Sun J, Hu Y, Fang Z, Bi Q and Chen Y 2015 Adsorption of Cu 2 + , Cd 2 + and Ni 2 + from aqueous single metal solutions on graphene oxide membranes **297** 251–60

[111] Tan P, Hu Y and Bi Q 2016 Competitive adsorption of Cu2+, Cd2+ and Ni2+ from an aqueous solution on graphene oxide membranes *Colloids Surfaces A Physicochem. Eng. Asp.* **509** 56–64

[112] Choi W, Choi J, Bang J and Lee J 2013 Layer-by-Layer Assembly of Graphene Oxide Nanosheets on Polyamide Membranes for Durable Reverse Osmosis Applications *ACS Appl. Mater. Interfaces* **5** (23) 12510–12519

[113] Zhang Y, Zhang S, Gao J and Chung T S 2016 Layer-by-layer construction of graphene oxide (GO) framework composite membranes for highly efficient heavy metal removal *J. Memb. Sci.* **515** 230–7

[114] Neelakandan S, Noel Jacob K, Kanagaraj P, Sabarathinam R M, Muthumeenal A and Nagendran A 2016 Effect of sulfonated graphene oxide on the performance enhancement of acid-base composite membranes for direct methanol fuel cells *RSC Adv.* **6** 51599–608

[115] Oh K, Son B, Sanetuntikul J and Shanmugam S 2017 Polyoxometalate decorated graphene oxide/sulfonated poly(arylene ether ketone) block copolymer composite membrane for proton exchange membrane fuel cell operating under low relative humidity *J. Memb. Sci.* **541** 386–92

[116] Jiang Z J, Jiang Z, Tian X, Luo L and Liu M 2017 Sulfonated Holey Graphene Oxide (SHGO) Filled Sulfonated Poly(ether ether ketone) Membrane: The Role of Holes in the SHGO in

IOP Conf. Series: Materials Science and Engineering **395** (2018) 012005 doi:10.1088/1757-899X/395/1/012005

Improving Its Performance as Proton Exchange Membrane for Direct Methanol Fuel Cells *ACS Appl. Mater. Interfaces* **9** 20046–56
[117] Kim B G, Han T H and Cho C G 2014 Sulfonated Graphene Oxide/Nafion Composite Membrane for Vanadium Redox Flow Battery *J. Nanosci. Nanotechnol.* **14** 9073–7
[118] Aziz M A, Oh K and Shanmugam S 2017 A sulfonated poly(arylene ether ketone)/polyoxometalate-graphene oxide composite: High ion selective membrane for all vanadium redox flow battery *Chem. Commun.* **53** 917–20
[119] Jiang Z, Zhao X and Manthiram A 2013 Sulfonated poly(ether ether ketone) membranes with sulfonated graphene oxide fillers for direct methanol fuel cells *Int. J. Hydrogen Energy* **38** 5875–84
[120] Lue S J ., Pai Y-L ., Shih C-M ., Wu M-C . and Lai S-M . 2015 Novel bilayer well-aligned Nafion/graphene oxide composite membranes prepared using spin coating method for direct liquid fuel cells *J. Memb. Sci.* **493** 212–23
[121] Sha Wang L, Nan Lai A, Xiao Lin C, Gen Zhang Q, Mei Zhu A and Lin Liu Q 2015 Orderly sandwich-shaped graphene oxide/Nafion composite membranes for direct methanol fuel cells *J. Memb. Sci.* **492** 58–66
[122] Bayer T, Selyanchyn R, Fujikawa S, Sasaki K and Lyth S M 2017 Spray-painted graphene oxide membrane fuel cells *J. Memb. Sci.* **541** 347–57
[123] Dai W, Shen Y, Li Z, Yu L, Xi J and Qiu X 2014 SPEEK/Graphene oxide nanocomposite membranes with superior cyclability for highly efficient vanadium redox flow battery *J. Mater. Chem. A* **2** 12423
[124] Chien H C, Tsai L D, Huang C P, Kang C Y, Lin J N and Chang F C 2013 Sulfonated graphene oxide/Nafion composite membranes for high-performance direct methanol fuel cells *Int. J. Hydrogen Energy* **38** 13792–801
[125] Bayer T, Bishop S R, Nishihara M, Sasaki K and Lyth S M 2014 Characterization of a graphene oxide membrane fuel cell *J. Power Sources* **272** 239–47
[126] Beydaghi H, Javanbakht M and Kowsari E 2014 Synthesis and characterization of poly(vinyl alcohol)/Sulfonated graphene oxide nanocomposite membranes for use in proton exchange membrane fuel cells (PEMFCs) *Ind. Eng. Chem. Res.* **53** 16621–32
[127] Kumar R, Mamlouk M and Scott K 2014 Sulfonated polyether ether ketone – sulfonated graphene oxide composite membranes for polymer electrolyte fuel cells *RSC Adv.* **4** 617
[128] Yin Y, Wang H, Cao L, Li Z, Li Z, Gang M, Wang C, Wu H, Jiang Z and Zhang P 2016 Sulfonated poly(ether ether ketone)-based hybrid membranes containing graphene oxide with acid-base pairs for direct methanol fuel cells *Electrochim. Acta* **203** 178–88

Acknowledgments

Financial assistance for this work has been provided by Institut Teknologi Bandung under a scheme 'Program Penelitian, Pengabdian, kepada Masyarakat, dan Inovasi (P3MI)'.

IOP Conf. Series: Materials Science and Engineering **395** (2018) 012006 doi:10.1088/1757-899X/395/1/012006

Influence of Steel Strip Temperature in Formation Zinc Dross During Process Production Using Continuous Galvanizing Line (CGL) Machine

R S Rachmat[1,2,*], M Ogawa[3]

[1]President University, Jalan Ki Hajar Dewantara, Cikarang Utara, Bekasi, Jawa Barat, Indonesia 17550
[2]Fumira, Jalan Jawa, Cikarang Barat, Bekasi, Jawa Barat, Indonesia 17520
[3]Nippon Steel Sumitomo Metals Engineering, 46-59, Nakabaru, Tobata-ku, Kitakyushu-shi, Fukuoka, Japan

*rudi.sr@president.ac.id

Abstract. To control the occurrence of zinc dross during galvanizing process using production Continuous Galvanizing Line (CGL) machine, it is necessary to control steel strip temperature before entry into zinc bath. The Steel strip temperature controls were variable at 320°C and 400°C and zinc bath temperature was fix at temperature 460°C. At temperature steel strips of 320°C, zinc bath was checked using spectrometry, the results of Fe content 0.1293% and Al 0.0759%, because the solubility of Al is lower than Fe, the dross occurs become bottom dross. If the zinc dross change become bottom dross, the bottom dross could decrease quality of product and life time of zinc pot, because bottom dross settled at the floor. Bottom dross properties very hard and absorb heating, then at that position bottom dross make hot spot in the floor, finally zinc pot leakage. At temperature steel strip 400°C, the results content of Al 0.1667% and Fe 0.1217%, due to high temperature steel strip make solubility of Al become higher than Fe, therefore float at the zinc bath become top dross.

1. Introduction

Hot- dip galvanizing is a method of zinc coating by dipping steel into zinc liquid, the purpose this process to improve corrosion resistance. There are two ways to coat the steel with liquid zinc, which steel dipping with high-speed into the liquid zinc, we know is called "continuous" galvanized line (CGL), and the next is dipping large and small steel into liquid zinc called batches process.

Both galvanizing processes always produce dross or impurities, this dross occurs due to the intermetallic reaction between iron with liquid zinc in zinc bath [1,2]. The dross formation rate in liquid zinc depends on steel strip temperature, iron and aluminium solubility in zinc bath [3].

This dross when floating above the surface of the liquid zinc become top dross, this is due to the low iron content in the zinc bath, but the high aluminium content [4], Top dross is intermetallic Fe_2Al_5Zn (η)[5], and affecting the quality of the product. And if the dross floats in liquid zinc, this is due to its lower aluminium content and high iron content then it becomes the bottom dross $FeZn_{10}$ (δ)[6,7]. This bottom dross when settled and become hard at zinc bath called hard zinc $FeZn_{13}$(ζ), Hard zinc will absorb heat then become hot spots on the walls make steel wall of zinc pot become hole and eventually leaking.

IOP Conf. Series: Materials Science and Engineering **395** (2018) 012006 doi:10.1088/1757-899X/395/1/012006

The purpose of this study is to investigate the effect of steel strip temperature before entry to zinc bath to the formation of top dross and bottom dross, and support by microstructures and chemical composition.

2. Materials and Methods

For this study directly using Continuous Galvanizing Line (CGL) production machine with the aim to be in accordance with the real industry. The Continuous Galvanizing Line CGL Production Machine used the NOF (Non-Oxidizing Furnace) system, the steel strip inside the furnace does not react with oxygen. Line speed of the steel strip entry into the zinc bath is 150 MPM and the zinc bath temperature was about 460°C. Scheme zinc bath galvanizing process, could be seen in 'Figure 1'.

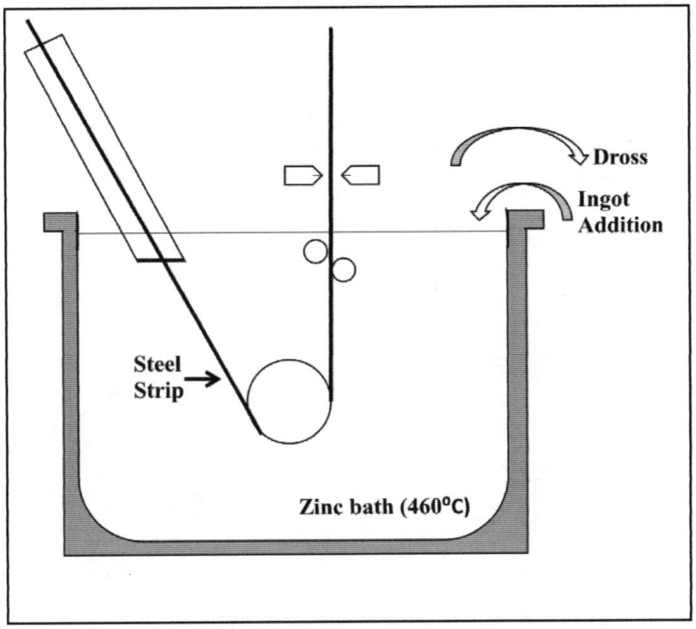

Figure 1. Zinc bath schematic on galvanizing process.

Steel strips were used Cold Roll Coil (CRC) as standard JIS G 3141 SPCC - 1D or 1B. And zinc ingot use CGG type. The chemical composition of zinc ingot could be seen in table 1. Production process parameters for line speed of the steel strip, the chemical composition of steel strip and chemical composition of zinc ingot were same, but the steel strip temperature parameters before entry into zinc bath using two variation temperatures with 320°C and 400°C.

Method for take bottom dross at the bottom position in pot is using steel stick, after the sharp of position, directly bottom dross take out using small pickup then poured into steel mould to make sample for spectrometry test.

Table 1. Chemical composition of zinc ingot (mass%).

Al	Cd	Cu	Fe	Pb	Sn	Zn
0.389 - 0.4022	<0.0001	<0.0006	0.0010 - 0.0015	<0.0025	<0.0001	Ball

3. Data and Analysis

In direct production process, steel strip temperature before entry into zinc bath was 320°C, after production for 7 days, and tonnage about 2100 Ton, from visual observations, in zinc bath has seen increased dross occur, and sticky on the strip product, this phenomenon would decrease the quality of the product. When checked floor of the zinc bath using stainless steel rods, it seems there were bottom dross settled at floor already. Then top dross float to the top, bottom dross settled on corner of the zinc pot, then examined chemical composition used spectrometry and microstructure.

Method for take sample for spectrometry test, the dross at the top position of zinc was take out using small pickup then poured into steel mould to make sample for spectrometry.

Table 2. Top dross chemical composition with steel strip temperature 320°C (Mass%).

Al	Cd	Cu	Fe	Pb	Si	Sn	Zn
0.1417	0.0001	0.0006	0.1251	0.0005	0.0001	0.0005	Ball

Result examination of top dross chemical composition with steel strip temperature 320°C shown in table 2. From the table shown the Al 0.1417% and Fe 0.1251% reacted. Al and Fe reacted in zinc bath become intermetallic compound $Fe_2Al_5Zn(\eta)$ where the Al content is more higher than Fe content, the floating dross called top dross. From the microstructural photograph, shown the Zn matrix and beta phase and there were found small amount of dirt spot, that was Fe.

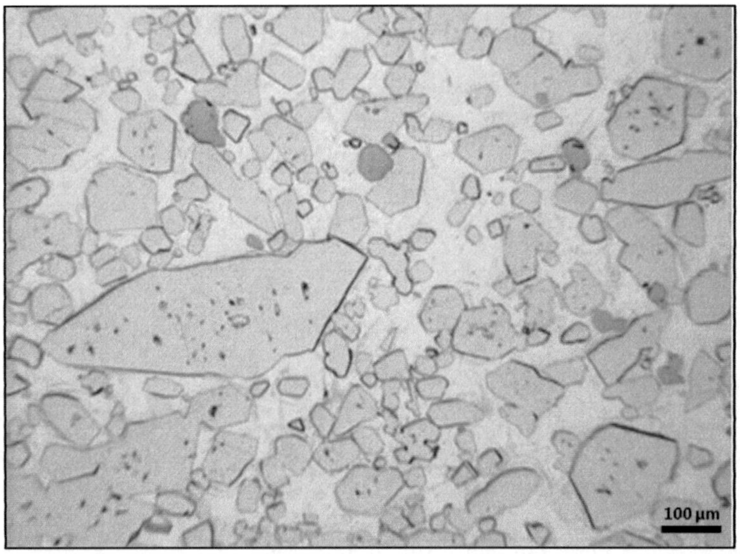

Figure 2. Top dross microstructure with steel strip temperature 320°C.

Table 3. Bottom dross chemical composition with steel strip temperature 320°C (% Mass).

Al	Cd	Cu	Fe	Pb	Si	Sn	Zn
0.0759	0.0001	0.0006	0.1293	0.0005	0.0001	0.0005	Ball

In table 3 Shown bottom dross chemical composition with steel strip temperature 320°C, from examination bottom dross chemical composition with steel strip temperature 320°C, entry into zinc bath at a temperature of 460°C, heat of steel strip was absorbed by zinc bath, made Al solubility in zinc block more slightly as in table 3 shown Al 0.0759% and Fe 0.1293%, because the Al content was less then Fe content, when Al and Fe reacted with zinc in zinc bath, become intermetallic zinc $FeZn_{10}(\delta)$, but specific gravity heavier than zinc, intermetallic dropped to down and settled in zinc pot to bottom become $FeZn_{13}(\zeta)$.
'Figure 3' shown the bottom dross microstructure with steel strip temperature of 320°C, shown the Zn matrix and beta phase, shown many dirt spot, that were Fe impurities due to too much solubility Fe.

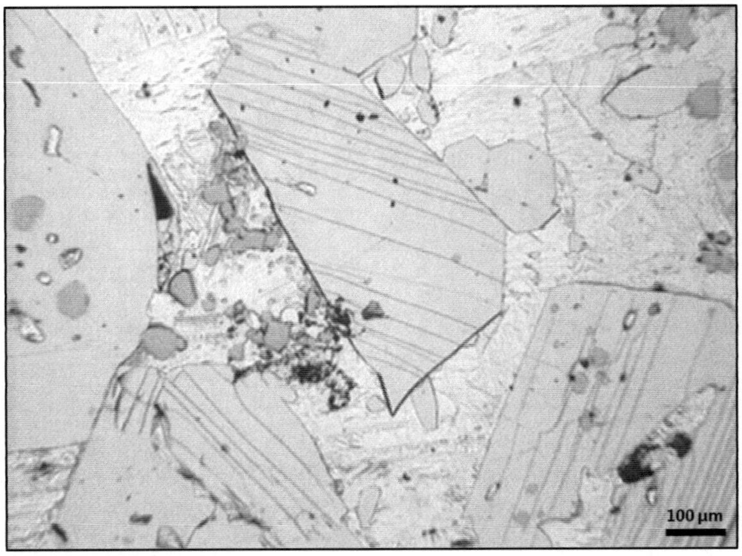

Figure 3. The bottom dross microstructure with steel strip temperature of 320°C.

Steel strip temperature 320°C before entry into zinc bath, many top dross and bottom dross happened, which greatly affects to quality. To increase quality of product, temperature steel strip was increased to 400°C, after 2 days production, tonnage of products about 600 tons, from visual inspection, the dross is started slightly decreased, then checked also by stainless steel rods, Top dross only occurred, bottom dross did not exist. After 7 days or tonnage of product about 2100 Ton The chemical composition and microstructure were checked, with steel strip temperature 400°C has seen in table 4. Shown Al content of 0.1667%, and Fe content of 0.1217%, Al when reacted with Fe in zinc bath, due to steel stripe temperature 400°C then the solubility of Al is more and then Fe solubility. Therefore, dross occur more less, consequently the top dross then floats more little.

Table 4. Top dross chemical composition with 400°C strip steel temperature.

Al	Cd	Cu	Fe	Pb	Si	Sn	Zn
0.1667	0.0001	0.0006	0.1217	0.005	0.001	0.005	Ball

'Figure 4'. Shown top dross microstructure with 400°C strip steel temperature. The results of examination microstructure of top dross with steel strip temperature 400°C, shown Zn matrix and beta phase, and shown more cleaner because solubility Fe decreased. Steel strips of temperature 400°C entry into zinc bath with zinc bath temperature of 460°C heat absorption only slightly, solubility of Al in zinc ingot increased, when Al reacted with Fe and Zn in zinc bath, the Al bonded Fe become $Fe_2Al_5Zn(\eta)$ which becomes dross and floats upward becomes top dross.

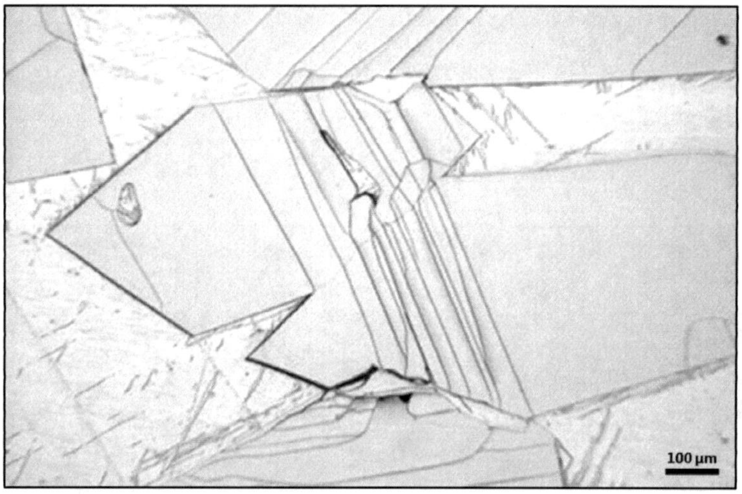

Figure 4. Top dross microstructure with 400°C strip steel temperature.

4. Conclusion

In this study, Influence of strip temperature in the formation of zinc dross by using two-temperature variations for steel strips were 320°C and 400°C, directly using production Continuous Galvanizing Line (CGL) machine, with zinc bath temperature of 460°C.

The conclusion as follows:

1. Both steel strip temperature 320°C and 400°C the Fe content in zinc bath were 0.12% - 0.13%.
2. Strip steel temperatures of 320°C, there are many top dross and bottom dross, because the solubility of Al was less in zinc bath while the solubility of Fe was higher.
3. Steel strip temperature of 400°C, the solubility of Al becomes higher, then only top dross occurred.

5. References

[1] J. Trpcevska, B. Hluchanova, T. Vindt, W. Zorawski, D.Jakubeczyova 2010 *Acta Metallurgical Slovaca* **16** 151 – 156

[2] Xingbo Liu, Ever Barbero, Jing Xu, Matthew Burris, Vinod Sikka, 2005 *Metallurgical and Materials Transaction A* **36**A

[3] B.N.Roy, Chandra Kamal Gupta 2016 *International Journal of Technical Research and Application* **4** 148 - 152

[4] Pranay Sawaitul, S.A. Chowriwar, I.P. Lade 2012 *International Journal of Emerging Technology and Advanced Engineering* **2** 45 - 51

[5] Jinichiro Nakano, Gary R. Purdy, V. Malakhov 2007 *Galvatech*'07, 135 – 140

[6] Sirli Sepper, Valdek Mikli 2016 *Proceeedings of the Estonian Academy of Science,* Material Engineering 159 – 165

[7] Ke Zhang, Nai-Yong Tan, Frank E. Goodwin, Scott Sexton, 2007 *J.Mater Sci* **23** 120 - 128

IOP Conf. Series: Materials Science and Engineering **395** (2018) 012007 doi:10.1088/1757-899X/395/1/012007

Critical Current Density, J_c Improvement of BSCCO Superconductor Tapes with Nano Co_3O_4 Addition

A N Jannah[1,*] and H Abdullah[2]

[1]Universiti Teknologi MARA, Negeri Sembilan Branch, Kuala Pilah Campus, Pekan Parit Tinggi, 72000, Kuala Pilah, Negeri Sembilan, Malaysia.
[2]Department of Electrical, Electronic and Systems Engineering, Universiti Kebangsaan Malaysia, 43600 UKM Bangi, Malaysia

Abstract. The $(Bi_{1.6}Pb_{0.4})Sr_2Ca_2Cu_3O_y(Co_3O_4)_x$ superconductor with addition of 0.00 - 0.05 weight per cent of Co_3O_4 nanoparticle with average size around 10 ± 4 nm was prepared using the co-precipitation method. Ag-sheathed Bi-2223 tapes with the highest critical current density, J_c of bulk $(Bi_{1.6}Pb_{0.4})Sr_2Ca_2Cu_3O_y(Co_3O_4)_x$ were fabricated using the powder-in-tube (PIT) method. The tapes were heat-treated at 845°C for 50 h. The structure, microstructure and transport critical current density (J_c) of the tapes were conducted using X-Ray diffraction, scanning electron microscopy and four point-probe respectively. The enhancement of critical current density, J_c through the addition of Co_3O_4 (10 nm) nanoparticles is reported. Our results show that all nanoparticles added samples for bulk and tape showed higher J_c compared to the non-added samples. The increase in J_c can be explained as the increase of the flux pinning strength by nanoparticles of Co_3O_4 that can act as effective pinning centres or a boost leading to enhancement of J_c in the Bi-2223 system.

1. Introduction

Superconducting BSCCO tape is one of the best materials for industrial and new promising product for high-current applications such as power transmission cables, high-efficiency generators, transformers and motors, current fault limiters, superconducting magnetic technology and other devices. Some of the important features of superconducting tape are very high currents carrying capability and negligible electric losses. The application of high-temperature superconductors requires the manufacture of longlength HTS tapes with high critical current density, J_c. [1-10]. The precursor powder, the mechanical processing parameters and the heat-treatment strategy are of great importance to the current leading potential of the final tape. To achieve the best results in critical current density, it is necessary to optimize several variables such as the starting powder composition, its phase distribution and particle size. Substitution of nanosized particles into them has been suggested as a novel approach to improve the transport capabilities of BSCCO superconductors [2-19]. These particles work as extra point-like pinning centres and pin the flux lines under applied magnetic fields. The pinning potential is one of the basic parameters that define the performance of a superconducting material and generally, the critical current density will increase when magnetic particles with characteristic size L is larger than the coherence length, ξ but smaller than the penetration depth, λ [17]. In this work, we report the effect of 10 nm sized Co_3O_4 addition (0 – 0.05 wt %) on the transport properties of bulk and (Bi,Pb)-2223/Ag tapes to study their effectiveness in enhancing J_c.

2. Experimental details

The $(Bi_{1.6}Pb_{0.4})Sr_2Ca_2Cu_3O_y(Co_3O_4)_x$ (x = 0.00 to 0.05 wt%) compound with and without Co_3O_4 were synthesized by using the co-precipitation method [20]. Commercial chemical precursor powders of Bi_2O_3, PbO, $SrCO_3$, $CaCO_3$ and CuO were used to produce $(Bi_{1.6}Pb_{0.4})Sr_2Ca_2Cu_3O_y$. The average particle size of Co_3O_4 that was added to the precursor powder is 10 nm. The additional amount of Co_3O_4 represents 0 to 0.05 wt % of the total mass of the sample. The weight percentage that gave the highest transport critical current density for the series was determined and loaded into a Ag tube with outer diameter of 6.03 mm and inner diameter of 4.43 mm. The assemblies were compacted by drawing and rolling to a final thickness ~0.20 mm. After being cut and uniaxially pressed into tapes, the samples were submitted to heat treatment at 845°C in air for 50 h, and furnace cooling to room temperature. The samples were submitted to the four-probe standard test for the critical temperature and current density characterization. The measurements of the transport critical current density (J_c) between 30 K to 77 K was estimated considering the established 1 μV/cm criterion by the four-point probe method. The surface morphology was evaluated by scanning electron microscopy (SEM) with x3.00 K and the distribution of nano Co_3O_4 in the sample and the presence of the different phases was analysed by using a Philips energy dispersive X-ray analyzer (EDX) model PV99 with CuKα radiation between 2Θ = 3° - 60°. The size of nano Co_3O_4 used in this work was determined using a Philips transmission electron microscope (TEM- model CM12) with x100000 magnitude.

3. Results and discussion

Figure 1 shows TEM micrographs of the Co_3O_4 nanoparticles used in this work with average size around 10 ± 4 nm. Figure 2 and 3 shows the critical current density, J_c versus Co_3O_4 content in bulk $Bi_{1.6}Pb_{0.4}Sr_2Ca_2Cu_3O_{10}$ at 30 K and 77 K. The highest critical current density, J_c was observed in the x = 0.03 wt% Co_3O_4 (10 nm) sample. Fig 4 shows the temperature dependence of J_c values at various temperatures in zero magnetic fields for tapes sample for $Bi_{1.6}Pb_{0.4}Sr_2Ca_2Cu_3O_{10}$, with 0.03 wt% of Co_3O_4 (10 nm). J_c decreased with increasing temperature as the consequence of thermal activated flux creep. Table 1 shows the J_c of non-added $Bi_{1.6}Pb_{0.4}Sr_2Ca_2Cu_3O_{10}$ tapes and with the Co_3O_4 nanoparticles addition at 30 and 77 K in zero fields. It is clear that the results show that J_c for added samples is higher than the non added tape sample with the maximum J_c at 30 K (15050 A cm^{-2}) and 77 K (3344 A cm^{-2}). J_c at 77 K for added sample is approximately three times higher than non-added sample. This results show that small amount of Co_3O_4 nanoparticles enhanced the transport critical current density in bulk and tape samples. The graph in Figure 3 shows the J_c values as a function of amount of Co_3O_4 additions. Co_3O_4 addition increased the J_c for 0.01 wt%, 0.02 wt% and 0.03 wt%. These results suggest that low concentration of Co_3O_4 contributed to the pinning properties and lowered the porosity levels in the structure. However, 0.04 wt% and 0.05 wt% addition decreased the J_c may be due to the accumulation of Co_3O_4 in the grain boundaries. Figure 5 shows the XRD patterns of tape sample for non added $Bi_{1.6}Pb_{0.4}Sr_2Ca_2Cu_3O_{10}$, and added sample with 0.03 wt% of Co_3O_4 (10 nm). X-ray diffraction patterns showed the added sample have quite similar lattice parameters compared with non-added sample. Bi2223 phase is dominant for while the Bi2212 is the minority phases in both samples. More Bi2223 phase formation was achieved with added sample. Here, H and L represent the diffraction peaks of high T_c phase (Bi2223) and low T_c phase (Bi2212) respectively obtained from the crystallographic planes. Co_3O_4 peaks were not observed by XRD analysis due to very low concentration and x values are too small to be detected. These results indicated that small amount of nano Co_3O_4 does not change the lattice parameters but this suggests that 0.03 wt% of Co_3O_4 nanoparticles addition to can accelerates the phase transformation from Bi2212 to Bi2223. It is also seen that the peak intensities of Bi2223 phase becomes sharper with Co_3O_4 addition. Scanning electron micrographs images of the tape sample for non-added $Bi_{1.6}Pb_{0.4}Sr_2Ca_2Cu_3O_{10}$, and added sample with 0.03 wt% of Co_3O_4 (10 nm) were examined to investigate the surface morphology as shown in Figure 6. It is known that additive materials in superconducting materials may increased the critical current density behaving as a pinning center taking an account of physical properties of additive materials such as size, amount and distribution. Microstructural investigation showed that plate-like layered microstructure are observed in the non-added sample which is the typical structure of BSCCO system. The sample with nano Co_3O_4 show smaller grain size than the undoped sample. The grain morphology

IOP Conf. Series: Materials Science and Engineering **395** (2018) 012007 doi:10.1088/1757-899X/395/1/012007

of the added samples is more or less identical texture with decreased level of porosity that enhances the critical current density. Nanosized Co_3O_4 particles in the granular networks are believed to act as possible pinning sites in the polycrystalline BSCCO.

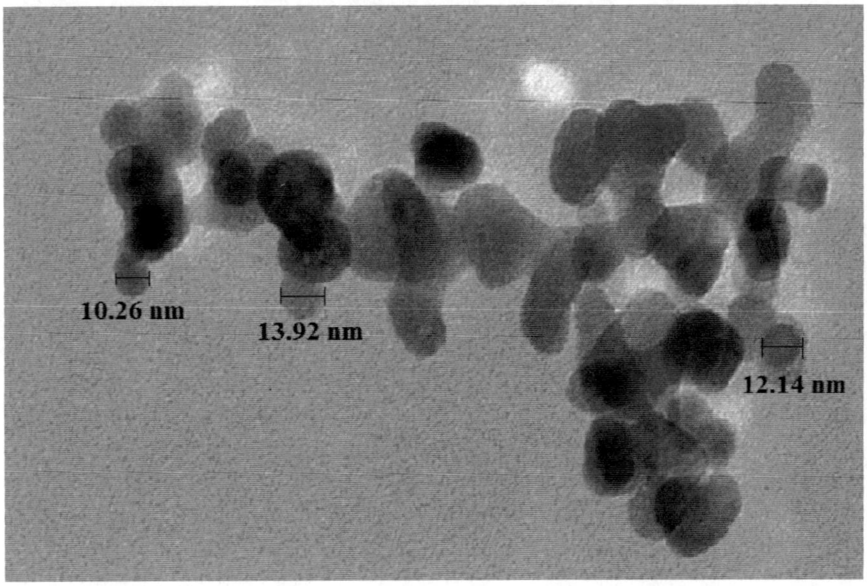

Figure 1. TEM micrograph of Co_3O_4 with average grain size of 10 ± 4 nm.

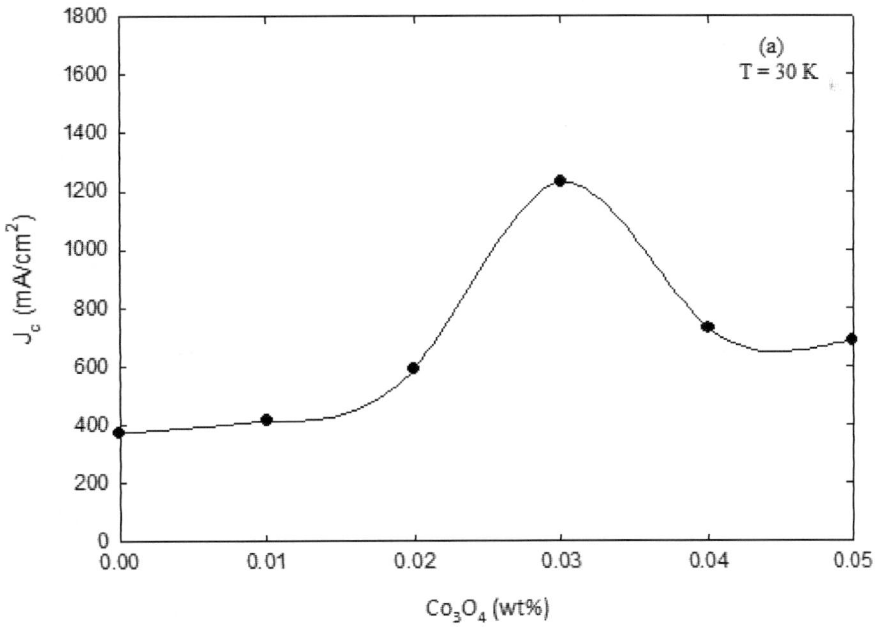

Figure 2. Critical current density, J_c versus Co_3O_4 (10 nm) content in bulk sample of $Bi_{1.6}Pb_{0.4}Sr_2Ca_2Cu_3O_{10}$ at 30 K.

IOP Conf. Series: Materials Science and Engineering **395** (2018) 012007 doi:10.1088/1757-899X/395/1/012007

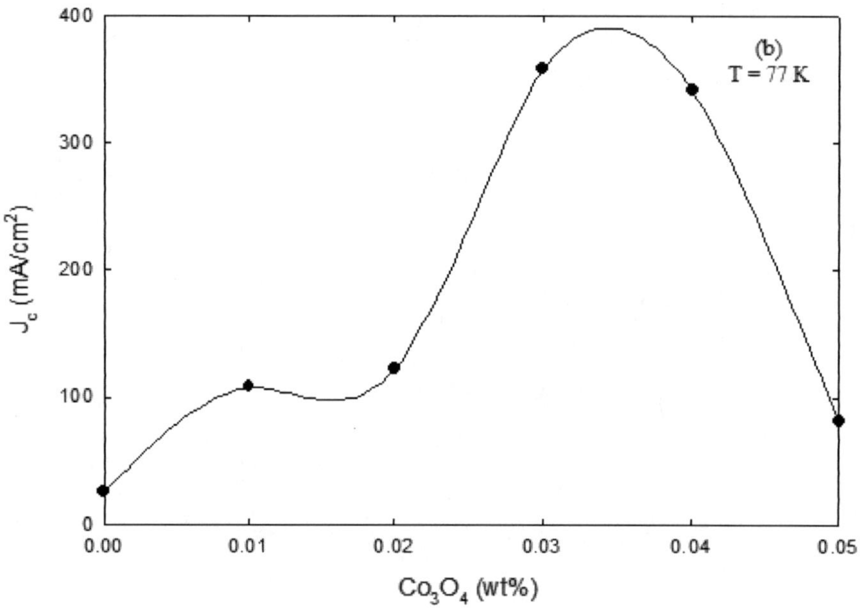

Figure 3. Critical current density, J_c versus Co_3O_4 (10 nm) content in bulk sample of $Bi_{1.6}Pb_{0.4}Sr_2Ca_2Cu_3O_{10}$ at 77 K.

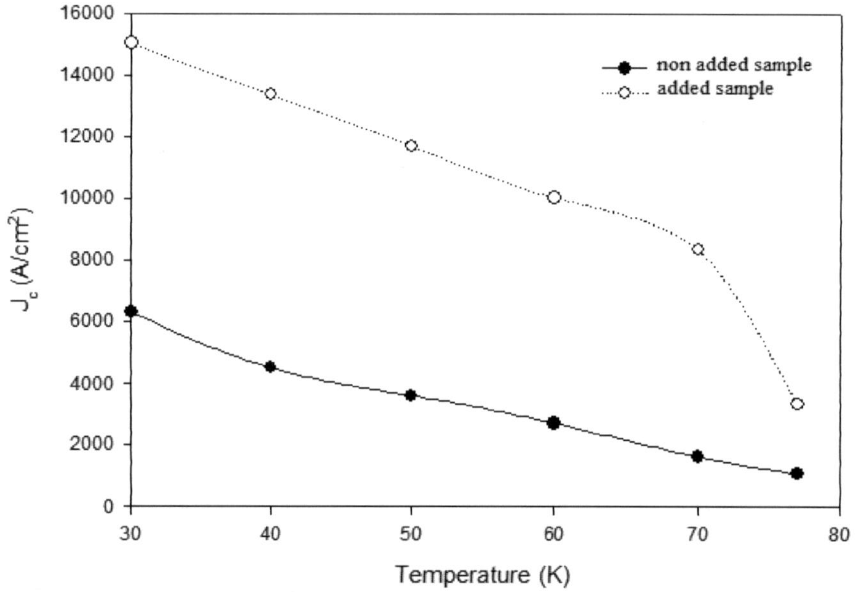

Figure 4. Temperature dependence of J_c in zero magnetic fields for tapes sample of $Bi_{1.6}Pb_{0.4}Sr_2Ca_2Cu_3O_{10}$ with 0.03 wt% of Co_3O_4 (10 nm) and without Co_3O_4.

IOP Conf. Series: Materials Science and Engineering **395** (2018) 012007 doi:10.1088/1757-899X/395/1/012007

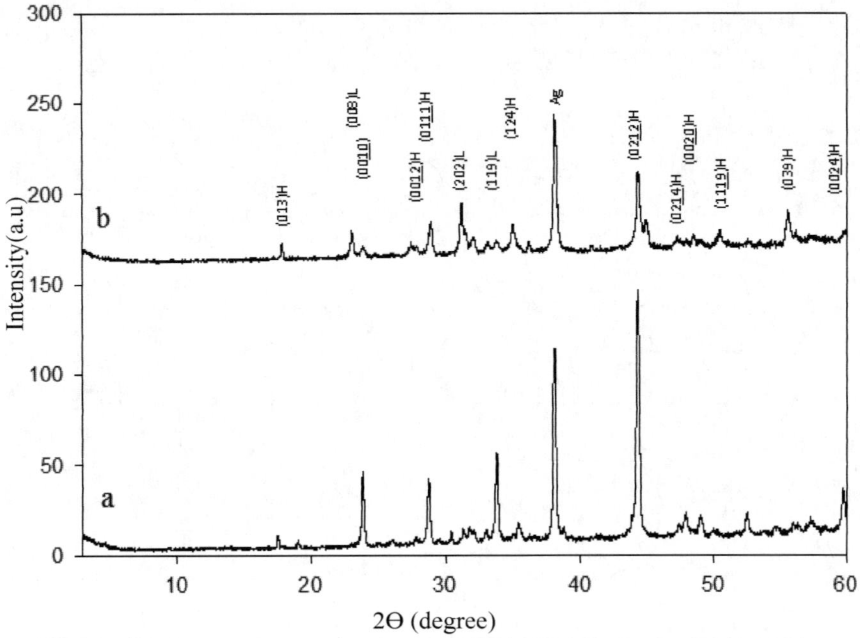

Figure 5. XRD patterns of tape sample for $Bi_{1.6}Pb_{0.4}Sr_2Ca_2Cu_3O_{10}$ a) without Co_3O_4 and b) added sample with 0.03 wt% of Co_3O_4 (10 nm). (H) denotes the high-T_c phase (Bi2223) and (L) denotes the low-T_c phase (Bi2212).

Table 1. Critical current density (J_c) at 30 and 77 K in zero fields for $Bi_{1.6}Pb_{0.4}Sr_2Ca_2Cu_3O_{10}$ tapes with and without the Co_3O_4 nanoparticles addition.

Composition	J_c at 30 K, A cm^{-2}	J_c at 77 K, A cm^{-2}
$Bi_{1.6}Pb_{0.4}Sr_2Ca_2Cu_3O_{10}$	6317	1083
$(Bi_{1.6}Pb_{0.4})Sr_2Ca_2Cu_3O_y(Co_3O_4)_{0.03}$ (10 nm)	15050	3344

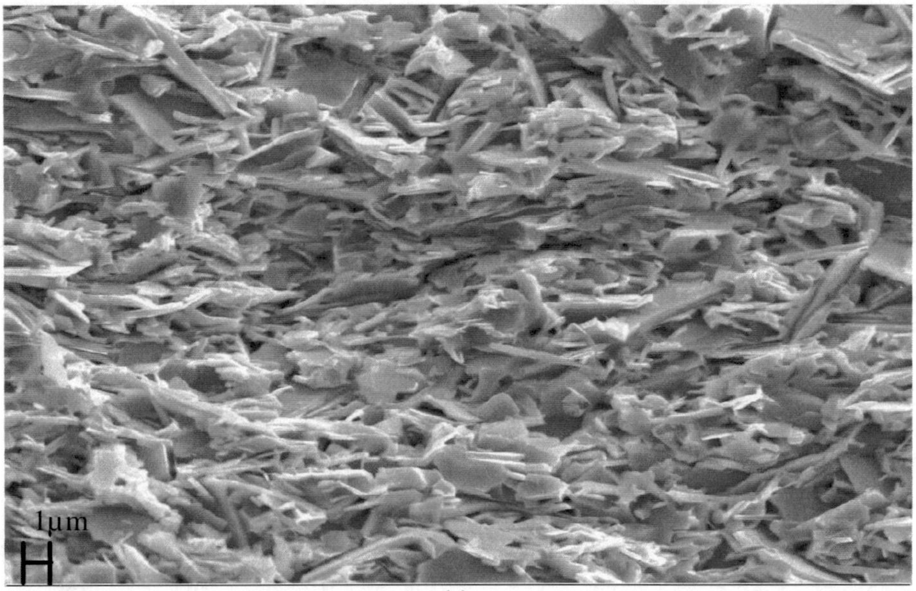

(a)

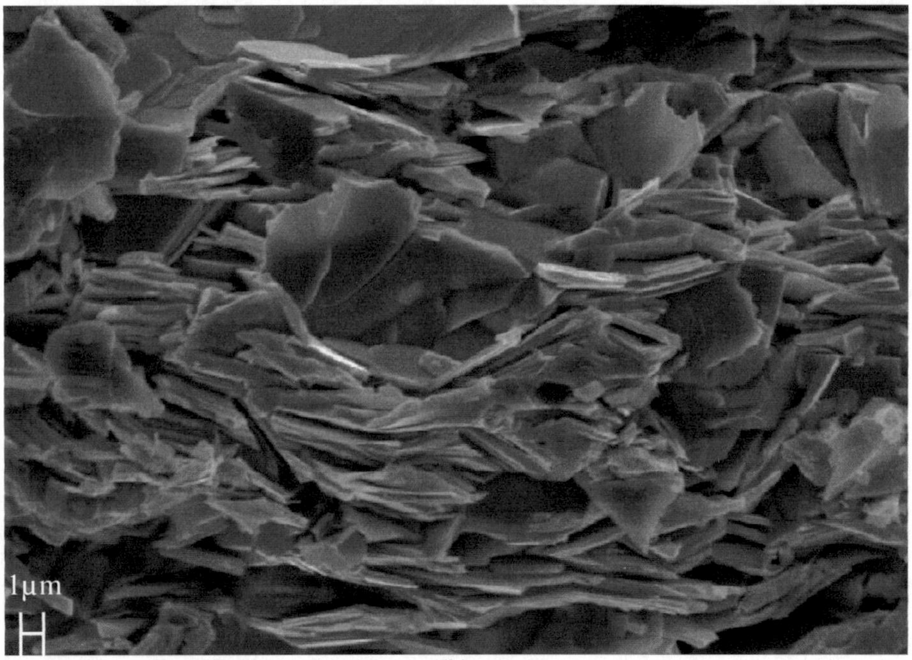

(b)

Figure 6. SEM micrographs of the tape samples for $Bi_{1.6}Pb_{0.4}Sr_2Ca_2Cu_3O_{10}$ added sample with 0.03 wt% of (a) Co_3O_4 (10 nm) and (b) without Co_3O_4.

IOP Conf. Series: Materials Science and Engineering **395** (2018) 012007 doi:10.1088/1757-899X/395/1/012007

4. Conclusions

In conclusion, the effect of nano Co_3O_4 with average size around ± 10 nm addition on flux pinning capability of the $(Bi_{1.6}Pb_{0.4})Sr_2Ca_2Cu_3O_y(Co_3O_4)_x$ superconductor and Ag-sheathed Bi-2223 tapes with the highest J_c of bulk $(Bi_{1.6}Pb_{0.4})Sr_2Ca_2Cu_3O_y(Co_3O_4)_x$ were investigated. The highest critical current density, J_c at 30 K and 77 K was observed in sample with $x = 0.03$ wt% Co_3O_4 (10 nm). Furthermore, all nanoparticles added samples showed higher J_c compared to the non-added samples. Tape samples (0.03 wt % of Co_3O_4) also shows high J_c compared to non-added tape sample. Further addition Co_3O_4, 0.04 wt% and 0.05 wt% adversely affected grain boundaries and caused a degradation in superconducting properties. By varying the composition of addition material, the superconducting critical current density, J_c is optimized and this study shows that nanoparticles can act as effective pinning centres leading to enhancement of J_c in the Bi-2223 system. Critical current density, J_c may be limited by the purity of the superconducting phase, the packing density, and the degrees of inter-grain connectivity and inter-grain alignment. In this work, the addition of Co_3O_4 to Bi-based ceramic has been found to have a beneficial effect of the critical current density, J_c associated to the flux pinning improvement. Generally, when the size of a pinning center is larger than the coherence length, ξ but smaller than the penetration depth, λ the critical current density will increase [12]. The average size of the Co_3O_4 used in this study is $L \sim 10$ nm and this satisfies the requirement $\xi < L < \lambda$ for the enhanced J_c in the system.

5. References

[1] Cenaa C R, Torsonib G B, Zadorosnyc L, Malmongec L F, Carvalhoc C L and Malmongec J A 2017 *Ceramics International* **43** 7663

[2] Öztürk H and Safran S 2018 *Journal of Alloys and Compounds* **731** 831

[3] Jiang J, Chai X Y, Polyanskii A A, Schwartzkopf L A, Larbalestier D C, Parrella R D, Li Q, Rupich M W and Riley Jr. G N 2001 *Superconductor Science and Technology* **14** 548

[4] Albiss B A, Obaidat I M, Gharaibeh M, Ghamlouche H and Obeidat S M, 2010 *Solid State Communications* **150** 33-34 1542

[5] Barik H K, Ghorai S K and Bhattacharya S 2000 *Journal of Materials Research* **15** (5)1076

[6] Chen X P, Grivel J C, Li M Y, Liu Q, Han Z, Andersen N H and Homeyer J 2010 *Materials Chemistry and Physics* **123** (2-3) 747

[7] Dou S X, Liu H K, Guo Y C and Sorrell C C 1992 *Superconductor Science and Technology* **5** (1S): S471

[8] Erdem M, Özturk Ö, Yucel E, Altintas S P, Varilci A, Terzioğlu C and Belenli I 2011 *Physica B: Condensed Matter* **406** (3): 705

[9] Ghazanfari N, Kılıç A, Gencer A and Özkan H, 2007 *Solid State Communications* **144**(5–6): 210

[10] Guilmeau E, Andrzejewski B and Noudem J G 2003 *Physica C* **387**: 382

[11] Gul I H, Amin F, Abbasi A, Anis-ur-Rehman Z and Maqsood A 2006 *Physica C: Superconductivity* **449** (2):139

[12] Jia Z Y, Tang H Z, Yang Q, Xing Y T, Wang Y Z and Qiao G W 2000 *Physica C: Superconductivity* **337** (1–4): 130

[13] Matsushita T, Suzuki A, Kishida T, Okuda M and Naito H 1994 *Superconductor Science and Technology* **7** (4): 222

[14] Ohba K, Horio T, Iwasaki K and Oota A 1992 *Superconductor Science and Technology* **5** (1S):S312

[15] Jannah A N, Halim S A and Abdullah H2009 *European Journal of Scientific Research* **29** 438

[16] Osabe G 2016 *Journal of the Japan Institute of Metals* **80** (7) 396

[17] Pu M H, Song W H, Zhao B, Wu X C, Hu T, Sun Y P and Du J J 2001 *Superconductor Science and Technology* **14** 305

[18] Çelebi S and Düzgün İ 2009 *Superconductor Science and Technology* **22** 034018

[19] Lyuksyutov I F and Naugle D G 1999 *Modern Physics Letters B* **13** (15): 491

[20] Azhan H, Hawa J S, Azman K, Hidayah H N and Yusainee S Y S 2013 *Advanced Materials Research* **622** 177

Acknowledgments
This research was supported by the Ministry of Science of High Education under grant no. ERGS/1/2011/STG/UKM/01/25 and Universiti Kebangsaan Malaysia under grant no. UKM-DLP-2011-018.

IOP Conf. Series: Materials Science and Engineering **395** (2018) 012008 doi:10.1088/1757-899X/395/1/012008

Fabrication of Ag-Loaded TiO_2 Nanotube Arrays on Ti-6Al-7Nb for Dental Implants Application

A R Murnandityas[1], M Ibadurrohman[1], B M Bachtiar[2] and Slamet[1,3]

[1]Department of Chemical Engineering, Faculty of Engineering, Universitas Indonesia, Depok 16424
[2]Department of Oral Biology and Oral Science Research Center, Faculty of Dentistry, Universitas Indonesia, Jakarta 10430
[3]Author to whom any correspondence should be addressed

E-mail: slamet@che.ui.ac.id

Abstract. In the present study, Ag-loaded Titanium Nanotube Arrays (TNTAs) were fabricated on the surface of non-toxic dental material, Ti-6Al-7Nb. A layer of TNTAs were introduced on Ti-6Al-7Nb surface, hence extending its effective area and inducing antibacterial properties for biofilm inhibition. TNTAs were generated on the surface of Ti-6Al-7Nb by an anodization method in an electrolyte solution containing glycerol, water and NH_4F. This was followed by depositing Ag via a photo-assisted deposition method to obtain Ag-loaded TNTAs. Surface morphology and crystallinity were evaluated by FESEM-EDX and XRD, respectively. FESEM-EDX analysis showed that Ag-loaded TNTAs were formed and well aligned on the surface of Ti-6Al-7Nb, while XRD confirmed the presence of anatase and Ag crystals.

Keywords: Dental implants; Ti-6Al-7Nb; Silver; TNTAs

1. Introduction

Dental implantation is considered as reliable treatment option for teeth restoration, owing to its prominent features in terms of strength, effectiveness and durability. Its application, however, is by no means without side-effects. Biocompability of dental implant material and risk of peri-implantitis have become concern in dental implantation.

Ti-6Al-4V as titanium alloys is the most common material for dental implants, due to its preeminent mechanical properties. However, biocompability of Ti-6Al-4V has been called to account, due to its vanadium content that can cause local adverse tissue reactions and immunological responses [1, 2]. Potentially toxic elements such as vanadium, copper and tin should be avoided completely from biomedical material applications [3]. Therefore, vanadium-free titanium alloys, such as Ti-6Al-7Nb have been developed as implant materials [4].

Although application of Ti-6Al-7Nb may have solved toxicity problems, peri-implantitis has been identified as another risk in dental implants application. Peri-implantitis has been recognized to be inflicted by the formation of bacterial biofilms on dental implant surface [5]. In fact, Peri-implantitis was identified in 28% of subjects and 43% of implant sites [6].

Peri-implantitis is an inflammatory reaction with loss of supporting bone in the tissues surrounding an implant, hence causing failure in osseointegration process that leads to implant failure [7]. Bacterial biofilm formation occurs due to incapability of dental implants to inhibit bacterial biofilm growth [8]. Therefore, modifications of dental implant material are needed to prevent peri-implantitis.

One way to prevent peri-implantitis is through modification of dental implant material surface morphology, makes it more resistant to bacteria adhesion or to incorporate antibacterial agent on its surface [9]. Those two functions can be achieved by fabricating TiO_2 nanotube arrays (TNTAs) on the surface of Ti-6Al-7Nb as titanium alloys. TiO_2 (Titanium Oxide) is known for its photocatalytic properties and wide use in disinfection, and was reported to cause oxidative damage in absence of photooxidation [10, 11].

Formation of natural oxide Titanium layer (TiO_2) on the surface of Ti-6Al-7Nb provides implant osseointegration [12, 13]. Natural oxide Titanium layer also enhanced antibacterial activity, since TiO_2 is known for its photocatalytic properties and has been widely utilized in disinfection [14]. In addition, TiO_2 was also reported to cause oxidative damage in the absence of photooxidation [11].

However, natural oxide Titanium layer can be easily broken and does not perform adequate protective functions in the working condition of a biological environment [12,13,14]. Consequently, modification of the surface morphology is needed to induce implant antibacterial activity and osseointegration process. Synthesis of TiO_2 nanotube arrays (TNTAs) is one way to generate TiO_2 layer on the surface of Ti-6Al-7Nb as dental implant alloys. TiO_2 in nanotube arrays will maximize the spesific surface area for catalytic reaction in disinfection process [15].

TiO_2 nanotube arrays (TNTAs) is generated on Titanium-based dental implant surface, hence extending its effective area, which constitutes more sites for Ag occupation during Photo-assisted Deposition (PAD) process as well as more reactive sites for disinfection process. Silver (Ag) also has been utilized as an antibacterial agent with low toxicity in the human body [16]. Silver (Ag) may also enhance photocatalytic efficiency, by preventing recombination of electrons and holes produced upon photo-driven excitation [17].

In the presence of Ag, mechanism of biofilm adhesion inhibition is expected to be driven by the anti-bacterial features of Ag. However, it should be noted that anti-bacterial properties of Ag nanoparticles have not fully understood yet. According to Zheng, et al (2012), silver nanoparticles could damage the microbial cell-wall structure or cause cell separation, and inhibit the adhesive abilities by preventing gene expression. Also, the nanostructure of materials may also be effective in reducing related gene expression, possibly by damaging the cell wall, effecting cell separation, or altering the organism's adhesive ability [18].

Without the presence of Ag nanoparticles, the mechanism of biofilm inhibition is most likely driven by the reactive oxygen species (ROS) generated from photocatalytic activities. These species, which include superoxide anion radicals, hydroxyl radicals, hydrogen peroxide, or atomic oxygen radicals are expected to damage DNA, RNA, proteins, and lipids of the bacteria in aqueous media. This subsequently leads to bacterial cell malfunction or destruction of the cell wall, extracting cell liquids and causing dehydration [19].

Among all the methods for preparation of TiO_2 nanotube arrays, such as sol-gel [20, 21], hydrothermal [22, 23] and chemical vapor disposition, the electrochemical anodization [24-28] has the advantages of low cost, simple equipment, and easy process [29, 30]. On the other hand, this method has been recognized as an efficient approach to produce integrative, vertically and highly ordered nanotube arrays with controllable dimensions of nanotubes [26]. Hence, electrochemical anodization is suitable for creating nanotubular oxide layers for biological applications, as it also offers an excellent strength of layer adhesion as compared to those fabricated by other techniques [13, 15].

In the present work, a two fold strategy to produce advanced implant materials with capability of preventing peri-implantitis were implemented. Anodization of Ti-6Al-7Nb plate in a NH_4F containing electrolyte successfully generated a nanotubular layer of TiO_2 on the surface, which is followed by photo-assisted deposition of Ag nanoparticles. Further study about the modified samples capability to prevent biofilm is also essential. Influence of surface modification and concentration of Ag should be studied in order to optimized parameters in this research. Biofilm assay to measure the capability of the as-synthesized Ag-loaded TiO_2 nanotubes is currently in progress and will be reported in the future paper. This work was done to further develop potential applications of TiO_2 nanotubes on modification of Ti-6Al-7Nb with similar methods in previous study by Rafieerad et al (2015) [13]. Moreover, this study is expected to contribute to the advancement of highly functioning and biocompatible implant materials.

2. Methods

Ti-6Al-7Nb plate (200 x 110 x 6 mm^3, Baoji Chuangxin Metal Materials Co. Ltd China) were used as substrate. The plate was cut into 40 x 10 mm pieces. Prior to anodic oxidation, the Ti-6Al-7Nb samples were ground with abrasive SiC papers up to 1500 CW, and then wet-polished in a mixture of HF, HNO_3 and demineralized water. The samples were subsequently cleaned by sonication in water for 5 min. Anodization was carried out in a two-electrode electrochemical cell in which a platinum sheet and the samples were connected to cathode and anode electrodes, respectively. The anodization process was carried out using a direct current (DC) power source at a constant potential of 50 V for 2 h. Optimized electrolyte mixture of glycerol, demineralized water and NH_4F (with stirring at room temperature) were used. After anodization, all of the samples were washed with demineralized water, and then annealed at 500 C for 3 h in compressed air atmosphere to obtain crystalline phases. The annealed samples (TNTAs/Ag/0) were soaked in $AgNO_3$ solutions (with ethanol as solvent) with three different concentrations (0.05 M, 0.1 M and 0.15 M) and irradiated with UV light for 3 hours while stirred. These samples were denoted as TNTAs/Ag/0.05, TNTAs/Ag/0.1 and TNTAs/Ag/0.15, respectively. The obtained Ag-loaded TNTAs Ti-6Al-7Nb samples were washed thoroughly using DI water, and cleaned ultrasonically afterwards. Micro-imaging of the samples was accomplished by a Field Emission Scanning Electron Microscope (FESEM, JEOL JIB-4610F) operated at EHT of 15 kV. The microscope is also equipped with a Thermo Noran spectrometer for EDX elemental probing. X-ray diffraction patterns of the prepared samples were recorded using an X-ray Diffractometer (PANalytical Empyrean), in order to elucidate their crystallite properties.

3. Results and Discussions

3.1. Surface morphology of modified material

Anodization of Ti-6Al-7Nbplates in the condition presented in this report results in formation of a TiO_2 nanotube layer on the plate surface as depicted in the FESEM micrographs (Fig. 1a). The resultant nanotubular structure is vertically aligned and highly ordered across the surface, although variation in terms of tube height and diameter can be observed from 3D image in figure 1b. The average internal diameter of the nanotubes is *ca.* 130 nm with wall thickness of *ca.* 30 nm. A slight irregularity of the tube height might be originated from the original roughness of the plate or from two different phases (α+β) that might exist on the surface of Ti-6Al-7Nb. Nb and Al were added onto Ti to make an alloy with improved mechanical strength, considering the fact that Ti in α-phase has relatively low mechanical strength. Accordingly, the presence of Nb and Al assisted Ti to transform into its β-phase to some extent, resulting a stronger metallic structure. This observation is in agreement with that reported by Macak (2005) who also investigated the formation of TNTAs on the surface of Ti-6Al-7Nb in an Ethylene-Glycol-contained electrolyte [31].

Figure 1. Surface morphology of modified Ti-6Al-7Nb (a) top view (b) tilted view.

Fig. 2 showed the morphology of TNTAs layer after the deposition of Ag, revealing that silver nanoparticles (white spots) with particle size in the range of 40–60 nm were well dispersed in the superstructure of TNTAs. These particles are spread across the surface, situated on top of TNTAs, as well as on both the inner and outer sides of the tube walls (Fig. 2). These results imply that the presence of nanotubular TiO_2 on the surface of Ti-6Al-7Nb provides extensive sites for Ag particles to occupy during photo-assisted deposition. Ultimately, both factors, *i.e.* higher surface area and adequate amount of deposited Ag particles, are expected to play significant roles in bactericidal activities and inhibition of biofilm formation.

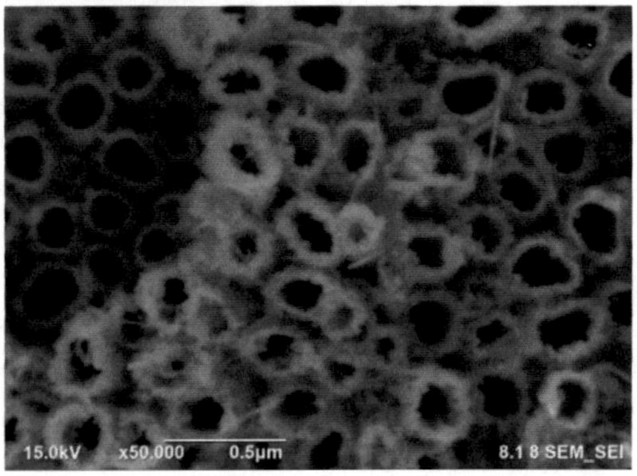

Figure 2. Ag-loaded TNTAs on modified Ti-6Al-7Nb.

Fig. 3 shows FESEM images of the modified samples, while fig. 4 shows the elemental (Ti, O and Ag) EDX mapping that were based on figure 2. From figure 3, addition of various concentration of Ag on the surface of TNTAs results in different distribution of Ag. Fig. 4 illustrates the dispersion of Ag particles, scanned across the surface of the TNTAs by energy dispersive X-ray (EDX) elemental mapping.

The results confirm that Ag particles were successfully deposited on TNTAs, with the intensity of Ag signal (yellow spots) being positively correlated to the concentration of $AgNO_3$ during PAD operation. It is also inferred from Figure 3 that the homogeneity of the particle dispersion deteriorates as the concentration become higher. Future tasks with regards to the operating conditions of PAD will be exerted in order to improve the uniformity of the resultant particle dispersion.

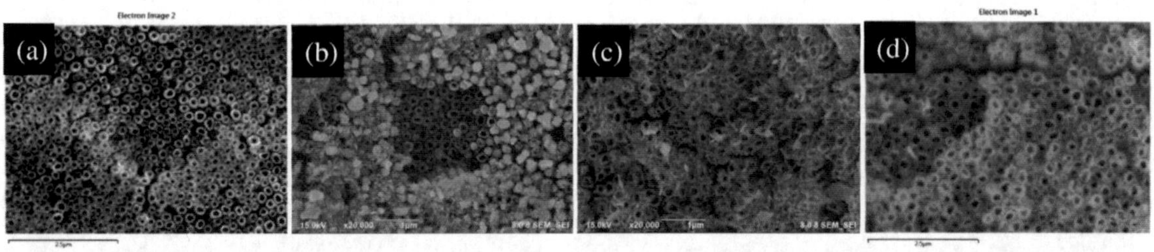

Figure 3. FESEM images of modified Ti-6Al-7Nb samples (a) TNTAs/Ag/0 (b) TNTAs/Ag/0.05 (c) TNTAs/Ag/0.1 (d) TNTAs/Ag/0.15.

Depositing Ag nanoparticles on a support with a nanotubular structure, as described in this report, is believed to offer advantageous features in the context of bacteria disinfection. Akhavan et al. [32] stipulated two key points that ensure the effectiveness of Ag as an anti-bacteria agent. Firstly, Ag-based materials should be loaded on a nanostructured support which provides a high surface-area-to-volume ratio (enormous active sites). By imposing higher interfacial area between Ag particles and their bacteria target, microbial activities are expected to be significantly enhanced without the need to increase the amount of deposited silver. Considering the fact that silver is a high-value metal, this strategy offers economic benefits in addition to technical advantages. Secondly, immobilization of Ag nanoparticles in a nanostructured and porous host would ensure their durability and long-term antimicrobial activities, because immobilized particles are more stable against washing which might otherwise give rise to material losses.

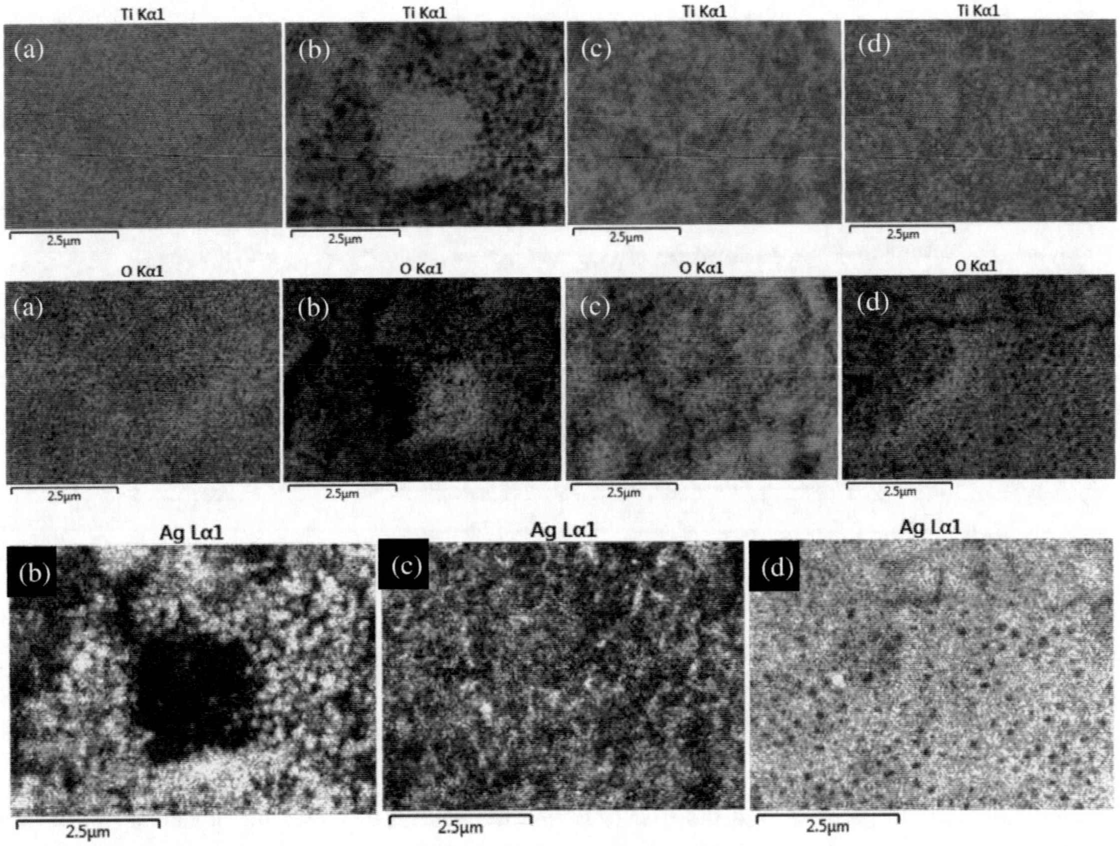

Figure 4. EDX mapping of Titanium (top), Oxygen (middle), and Silver (bottom) of modified Ti-6Al-7Nb samples (a) TNTAs/Ag/0 (b) TNTAs/Ag/0.05 (c) TNTAs/Ag/0.1 (d) TNTAs/Ag/0.15.

The XRD patterns of all samples are depicted in fig. 5. As shown in figure 5, the diffraction peaks appeared at 2θ = 25.4°, 37.86°, and 48.1° could be assigned to TiO_2 anatase (JCPDS#01-0711166). From XRD results, it can be concluded that TiO_2 anatase phase is obtained in all samples. Presence of TiO_2 rutile wasn't detected in the XRD profile, since rutile usually obtained via high temperature (greater than 800°C) calcination of anatase or in an acidic hydrothermal process.

Presence of Ag (JCPDS#04-0783) can be found where the diffraction peaks at 2θ value of 44.42°. The other diffraction peaks at about 2θ = 35°, 38.4°, 40.3° and 53.1° can be indexed to the titanium substrate (JCPDS#44-1294). From figure 5 modified sample of Ti-6Al-7Nb, with Ag-loading concentration of 0.15 M and 0.05 M shows peaks at 44.42°, while it is not detected in 0.1 M Ag-loaded TNTAs. These may occur due to Ag deposited in TNTAs/Ag/0.05 existed in small amount, as confirmed with EDX mapping analysis. Using Scherrer equation, the size of TiO_2 anatase crystalline can be obtained, with value 44.1 nm. The photocatalytic activity of the anatase form of TiO_2 is approximately 1.5 times higher than that of the rutile form. Accordingly, with XRD analysis confirms presence of anatase, photocatalytic activity may be enhanced.

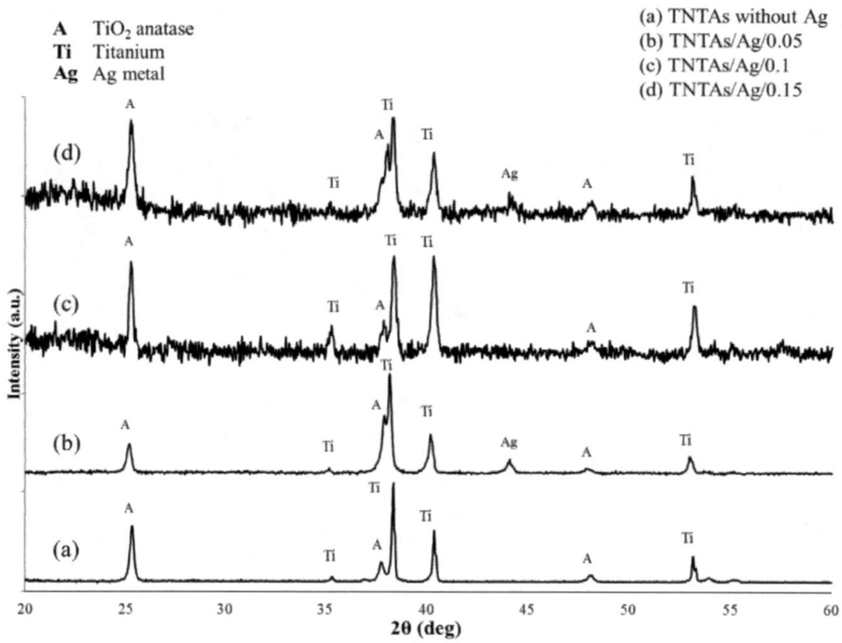

Figure 5. XRD patterns of modified Ti-6Al-7Nb samples (a) TNTAs/Ag/0 (b) TNTAs/Ag/0.05 (c) TNTAs/Ag/0.1 (d) TNTAs/Ag/0.15.

4. Conclusions

In this study, Ag-loaded TiO_2 nanotubes layer was successfully generated on Ti-6Al-7Nb surface with a regularly aligned nanotubular structure, hence extending its effective surface area. A photo-assisted deposition technique was evidently an effective way to deposit Ag nanoparticles on TiO_2 surface, as an electron trapper during activation and/or an antibacterial agent. Thus, some parameters of this research still need to be optimized in order to obtain the best strategy for the advancement of dental implant technology. In addition, further biofilm assay to measure capability of modified samples to inhibit bacterial biofilm growth is still in progress, and will be reported in a forthcoming paper.

5. References

[1] Okazaki Y, Gotoh E 2005 Comparison of metal release from various metallic biomaterials in vitro *Biomaterials* **26** 11-21

[2] Elias CN, Lima JHC, Valiev R, Meyers MA 2008 Biomedical applications of titanium and its alloys *Jom* **60** 46-9

[3] Dominguez COaA. Titanium as a Biomaterial for Implants. In: Fokter SK, editor. Recent Advances in Arthroplasty: CC BY 3.0 license; 2012. p. 149 - 62.

[4] Manfred F. Semlitsch HW, Robert M. Streicher and Rolf Sch6n 1992 Joint replacement components made of hot-forged and surface-treated Ti-6Al-7Nb alloy *Biomaterials* **13**

[5] Dhir S 2013 Biofilm and dental implant: The microbial link *J Indian Soc Periodontol* **17** 5-11

[6] Zitzmann NU, Berglundh T 2008 Definition and prevalence of peri-implant diseases *J Clin Periodontol* **35** 286-91

[7] Prathapachandran J, Suresh N 2012 Management of peri-implantitis *Dent Res J (Isfahan)* **9** 516-21

[8] Yoshinari M, Oda Y, Kato T, Okuda K 2001 Influence of surface modifications to titanium on antibacterial activity in vitro *Biomaterials* **22** 2043-8

[9] Necula BS, Fratila-Apachitei LE, Zaat SA, Apachitei I, Duszczyk J 2009 In vitro antibacterial activity of porous TiO_2-Ag composite layers against methicillin-resistant Staphylococcus aureus *Acta Biomater* **5** 3573-80

[10] Laxma Reddy PV, Kavitha B, Kumar Reddy PA, Kim KH 2017 TiO_2-based photocatalytic disinfection of microbes in aqueous media: A review *Environ Res* **154** 296-303
[11] Gurr JR, Wang AS, Chen CH, Jan KY 2005 Ultrafine titanium dioxide particles in the absence of photoactivation can induce oxidative damage to human bronchial epithelial cells *Toxicology* **213** 66-73
[12] Albrektsson T, Branemark PI, Hansson HA, Lindstrom J 1981 Osseointegrated titanium implants. Requirements for ensuring a long-lasting, direct bone-to-implant anchorage in man *Acta Orthop Scand* **52** 155-70
[13] Rafieerad AR, Zalnezhad E, Bushroa AR, Hamouda AMS, Sarraf M, Nasiri-Tabrizi B 2015 Self-organized TiO2 nanotube layer on Ti–6Al–7Nb for biomedical application *Surface and Coatings Technology* **265** 24-31
[14] Foster HA, Ditta IB, Varghese S, Steele A 2011 Photocatalytic disinfection using titanium dioxide: spectrum and mechanism of antimicrobial activity *Appl Microbiol Biotechnol* **90** 1847-68
[15] Roy P, Berger S, Schmuki P 2011 TiO2 nanotubes: synthesis and applications *Angew Chem Int Ed Engl* **50** 2904-39
[16] Lansdown ABG 2006 Silver in Health Care: Antimicrobial Effects and Safety in Use *Curr Probl Dermatol* **33** 17-34
[17] Seery MK, George R, Floris P, Pillai SC 2007 Silver doped titanium dioxide nanomaterials for enhanced visible light photocatalysis *Journal of Photochemistry and Photobiology A: Chemistry* **189** 258-63
[18] Zheng Y, Li J, Liu X, Sun J 2012 Antimicrobial and osteogenic effect of Ag-implanted titanium with a nanostructured surface *Int J Nanomedicine* **7** 875-84
[19] Buser D, Mericske-Stern R, Bernard JP, Behneke A, Behneke N, Hirt HP, et al. 1997 Long-term evaluation of non-submerged ITI implants. Part 1: 8-year life table analysis of a prospective multi-center study with 2359 implants *Clin Oral Implants Res* **8** 161-72
[20] Maiyalagan T, Viswanathan B, V. Varadaraju U. Fabrication and characterization of uniform TiO_2 nanotube arrays by Sol-gel template method2006. 705-8 p.
[21] Kang T-S, Smith AP, Taylor BE, Durstock MF 2009 Fabrication of Highly-Ordered TiO_2 Nanotube Arrays and Their Use in Dye-Sensitized Solar Cells *Nano Letters* **9** 601-6
[22] Yu K-P, Yu W-Y, Kuo M-C, Liou Y-C, Chien S-H 2008 Pt/titania-nanotube: A potential catalyst for CO_2 adsorption and hydrogenation *Applied Catalysis B: Environmental* **84** 112-8
[23] Sreekantan S, Wei LC 2010 Study on the formation and photocatalytic activity of titanate nanotubes synthesized via hydrothermal method *Journal of Alloys and Compounds* **490** 436-42
[24] Hsieh C-T, Lai M-H, Pan C 2010 Synthesis and visible-light-derived photocatalysis of titania nanosphere stacking layers prepared by chemical vapor deposition *Journal of Chemical Technology & Biotechnology* **85** 1168-74
[25] Gong D, Grimes CA, Varghese OK, Hu W, Singh RS, Chen Z, et al. 2001 Titanium oxide nanotube arrays prepared by anodic oxidation *Journal of Materials Research* **16** 3331-4
[26] Rohani S, editor Synthesis of titania nanotube arrays by anodization2009: Citeseer.
[27] Macak JM, Schmuki P 2006 Anodic growth of self-organized anodic TiO_2 nanotubes in viscous electrolytes *Electrochimica Acta* **52** 1258-64
[28] Sreekantan S, Hazan R, Lockman Z 2009 Photoactivity of anatase–rutile TiO_2 nanotubes formed by anodization method *Thin Solid Films* **518** 16-21
[29] Hsiao P-T, Liou Y-J, Teng H 2011 Electron transport patterns in TiO_2 nanotube arrays based dye-sensitized solar cells under frontside and backside illuminations *The Journal of Physical Chemistry C* **115** 15018-24
[30] Hu J, Zhao L, Yang Y, Liao H, Wang S, Sun X, et al. 2014 TiO_2 nanotube arrays composite film as photoanode for high-efficiency dye-sensitized solar cell *International Journal of Photoenergy* **2014**

[31] Macak JM, Tsuchiya H, Taveira L, Ghicov A, Schmuki P 2005 Self-organized nanotubular oxide layers on Ti-6Al-7Nb and Ti-6Al-4V formed by anodization in NH_4F solutions *J Biomed Mater Res A* **75** 928-33
[32] Akhavan O, Ghaderi E 2009 Bactericidal effects of Ag nanoparticles immobilized on surface of SiO_2 thin film with high concentration *Current Applied Physics* **9** 1381-5

Acknowledgments
The authors would like to sincerely acknowledge Universitas Indonesia for providing the necessary facilities and resources for this research. This work has been supported by the Directorate of Research and Community Service in Universitas Indonesia (DRPM UI) for Hibah Publikasi Internasional Terindeks untuk Tugas Akhir (PITTA) 2017.

Characterization of Limestone in Pamekasan Madura Island as Raw Material for Producing Nano Precipitated Calcium Carbonate (NPCC)

F Munawaroh[1,*], L K Muharrami[2], T Triwikantoro[3] and Z Arifin[4]

[1,2]Natural Science Education Study Program, Faculty of Education, University of Trunojoyo Madura, Bangkalan 69162, Indonesia
[3,4]Physics Department, Faculty of Natural Science, Sepuluh Nopember Institute of Technology, Surabaya 60111, Indonesia

*Email: fatim@trunojoyo.ac.id

Abstract. Characterization of limestone in Pamekasan Madura island was investigated. So far, limestone at Madura island has low economic value as additional building material. Limestone was taken from three mining locations. The chemical composition of limestone was characterized by X-Ray Fluorescence (XRF), the crystalline phase was characterized by X-ray diffraction (XRD) and morphological investigation was analyzed by Scanning Electron Microscope (SEM). The results XRF tests are obtained limestone with very high purity from Blumbungan, Pamekasan, East Java with Ca content 99,25 wt%. Characterization by using XRD and SEM are showed 100% calcite phase with rhombohedral- cubic shape and a particle size of $CaCO_3$ ranges between 1.594 – 2.503 μm. In conclusion, limestones from Blumbungan are able to material for producing nano precipitated calcium carbonate (NPCC). Furthermore, economic value of limestone in Madura Island are able to advance.

Keyword: limestone, calcite, very high purity, nano precipitated calcium carbonate

1. Introduction

Madura island has a high mineral mining potential. One of them is a mineral potential of carbonate-based mineral such as limestone, calcite and dolomite. Based on the data Energi and Mineral Resources (ESDM) East Java [1], Madura island has reserve of limestone amount ±152.219.027.133,86 ton. Limestone at Madura island has low economic value ie additional building material.

Limestone based calcite will be high value if processed precipitated $CaCO_3$. Precipitated $CaCO_3$ (PCC) has widely application in many industries that is used as filler in rubber, paper making, printing ink, cosmetics, toothpaste, food industries, pharmaceutical industries, nutritional supplements, adhesives, sealant and cement [2- 9, 11].

Many researchers have been trying to get smaller PCC or Nano PCC from natural resources [2, 7-10]. PCC is able to provide the desired characteristics by controlling the parameters. Arifin *et al* have succeeded in producing precipitated calcium carbonate from lime solution with carbonation method [7-9]. Threveni *et al* have succeeded in producing NPCC with carbonation process through a closed loop reactor [10]. Ramasamy *et al* have succeeded in synthesizing precipitated $CaCO_3$ nanoparticles using different natural carbonate sources [2]. NPCC has unique and extraordinary properties, that is

IOP Conf. Series: Materials Science and Engineering **395** (2018) 012009 doi:10.1088/1757-899X/395/1/012009

related to the size of less than 100 nm or 100 x 10^{-9} m. The natural resources used are limestone with high purity [2, 7-10].
The research aims were to characterize limestone in Pamekasan Madura island, to obtained limestone with very high purity. It will be used as raw material for producing nano precipitated $CaCO_3$ (NPCC).

2. Methods
The sampling of limestone was taken from three mining locations in Pamekasan Madura island (East Java, Indonesia), that are Blumbungan, Larangan; Lesong Daya, Batu Marmar and Lebbek, Pakong. Potential data limestone is obtained from ESDM East Java [1]. The limestone was crushed and mashed with the mortar and sieved with 18 mesh and 35 mesh. The chemical compositions of limestone was characterized by X-Ray Fluorescence (XRF) PANalytical Minipal 4 Sulfur, the crystalline phase was characterized by X-ray diffraction (XRD) X-Pert MPD with Cu-K_α radiation (λ = 1,5406 Å) and scanned from 10° to 60°. Morphological investigation was analyzed by Scanning Electron microscope (SEM) FEI Inspect S50 X'pertPRO PANalytical.

3. Results and Discussion
The chemical composition of limestone from three mining locations in Pamekasan Madura island is shown in Tab.1. The highest calsium content in limestone from Blumbungan, Larangan District is 99.25 wt% and impurities have less content than 1%. Limestone from Lesong Daya contains 95.15 wt%, 4.4 wt% magnesium and <1 wt% other impurities. The lowest calcium content found in limestone from Lebbek is only 87.4 wt%,with impurities of magnesium 0.5 wt%, aluminium 1.2 wt% and iron 2.99 wt%.

Table 1. Chemical composition of limestone in Pamekasan Madura island.

Element (wt%)	Mining Location			Oxide (wt%)	Mining Location		
	Blumbungan, Larangan District	Lesong Daya, Batu Marmar District	Lebbek, Pakong District		Blumbungan, Larangan District	Lesong Daya, Batu Marmar District	Lebbek, Pakong District
Ca	99.25	95.15	87.4	CaO	99.06	93.11	82.23
Mg	-	4.4	0.5	MgO	-	6.5	0.99
Al	0.5	-	1.2	Al_2O_3	0.7	-	2
Si	-	-	5.71	SiO_2	-	-	9.96
S	-	-	0.36	SO_3	-	-	0.7
K	-	-	0.41	K_2O	-	-	0.35
Ti	-	-	0.32	TiO_2	-	-	0.34
Mn	-	-	0.042	MnO	-	-	0.034
Fe	0.17	0.17	2.99	Fe_2O_3	0.15	0.16	2.55
Cu	0.068	0.12	0.088	CuO	0.055	0.1	0.064
Sr	-	0.13	0.37	SrO	-	0.098	0.26
Zr	-	-	0.02	ZrO_2	-	-	0.03
In	-	-	0.6	In_2O_3	-	-	0.5
Ba	-	-	0.09	BaO	-	-	0.05
Br	0.007	-	-	Br	0.006	-	-

Figure 1 shows the XRD patterns of limestone from Pamekasan. The three limestones have different phases. Based on search and match, the diffraction peaks of limestone from Blumbungan are macthed with rhombohedral-cubic phase of calcite ($CaCO_3$) without impurity with reference to JCPDS: 00-005-0586, limestone from Lesong Daya are macthed with rhombohedral phase of ankerite ($Ca[Fe,Mg][CO_3]_2$) without impurity with reference to (JCPDS: 00-041-0586) and limestone from Lebbek are identified three phases i.e. calcite (JCPDS: 00-005-0586), ankerite (JCPDS: 00-041-0586) and trapezohedral phase of quartz (SiO_2) (JCPDS: 00-046-1045). The XRD pattern of limestone from Blumbungan has a calcite phase, similar to that studied by Arifin *et al* [7,9].

Semiquantitative analysis using MACTH software of limestone diffraction patterns is shown in Tab. 2. Based on table 2 , limestone from Blumbungan has 100% calcite phase, from Lesong Daya has 100% ankerite phase and from Lebbek has three phases ie 73.8 % calcite phase, 17.4 % quartz phase and 8.7 % ankerite phase.

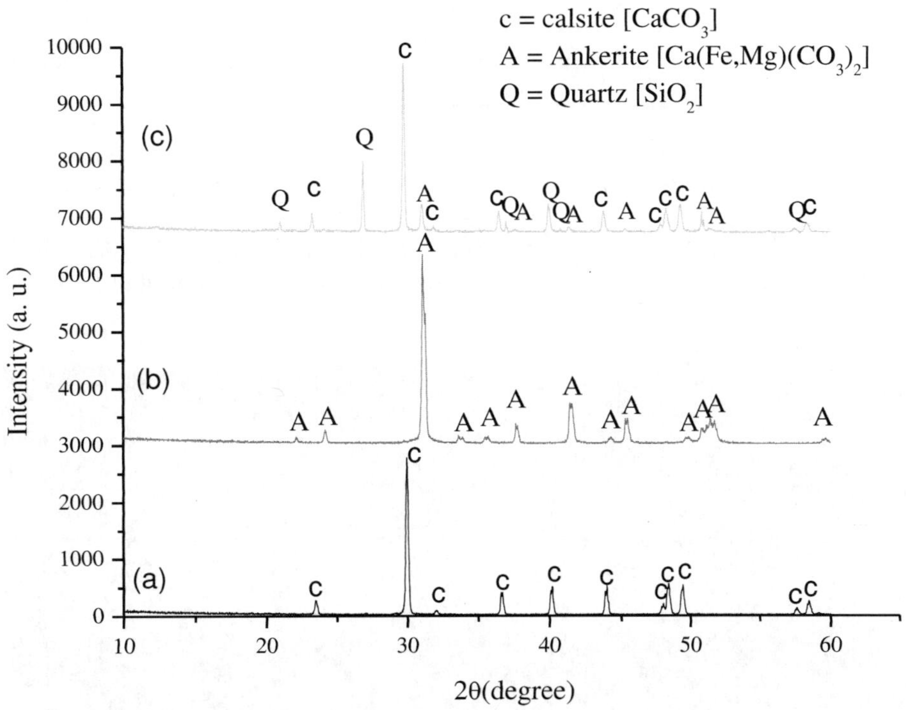

Figure 1. The XRD patterns of limestone (a) Blumbungan, Larangan; (b) Lesong Daya, Batu Marmar and (c) Lebbek, Pakong.

Table 2. Analysis semiquantitative of limestone in Pamekasan.

No	Mining Location	Chemical composition	
		Compound	Amount (%)
1	Blumbungan, Larangan	Calcite, $CaCO_3$	100
2	Lesong Daya, Batu Marmar	Ankerite, $Ca[Fe,Mg][CO_3]_2$	100
		Calcite, $CaCO_3$	73.8
3	Lebbek, Pakong	Quartz, SiO_2	17.4
		Ankerite, $Ca[Fe,Mg][CO_3]_2$	8.7

IOP Conf. Series: Materials Science and Engineering **395** (2018) 012009 doi:10.1088/1757-899X/395/1/012009

Table 3 shows limestone purity classification in Pamekasan Madura Island according to British Geological survey [11,12], which is based on XRF and XRD test results of the samples. The limestone from Blumbungan has a very high purity, while the limestone from Lesong Daya and Lebbek has a medium purity and impure, respectively. Therefore limestone from Blumbungan with very high purity has potential as raw material for producing nano calcium carbonate precipitated (NPCC), as previous researchers have done [2,7,9,10].

Table 3. Limestone purity classification in Pamekasan Madura Island according to *British Geological Survey* [11,12].

No	Mining Location	$CaCO_3$ wt% (XRD)	CaO wt% (XRF)	MgO wt% (XRF)	SiO2 wt% (XRF)	Fe2O3 wt% (XRF)	Purity Classification
1	Blumbungan, Larangan	100	99.06	-	-	0.15	Very high purity
2	Lesong Daya, Batu Marmar	-	93.11	6.5	-	0.16	Medium purity
3	Lebbek, Pakong	42.86	82.23	0.99	9.96	2.55	Impure

Crystal morphology of natural calcite from limestone Blumbungan with very high purity is shown in Fig. 2. Fig. 2(a) shows the $CaCO_3$ particles tend to agglomerate. Particle size of $CaCO_3$ ranges between 1.594 – 2.503 μm. The calcite phase has the rhombohedral-cubic shape(Fig. 2(b)). Similar results have been reported by Arifin et al [7,9], where the limestone from Tuban-East Java also has a rhombohedral cubic shape but with a size of 6 times larger than what we reported i.e. 15 μm.

Figure 2. Distribution of $CaCO_3$ particle limestone Blumbungan (a) magnification of 5000x and (b) magnification of 20000x.

4. Conclusions

Limestone from Blumbungan Pamekasan has very high purity with Ca content 99.25 wt% and pure calcite phase (100 wt% $CaCO_3$). Morphology of calcite is the cubic shape with a particle size of $CaCO_3$ ranges between 1.594 – 2.503 μm. Limestone from Blumbungan can be used as raw material for producing nano precipitated calcium carbonate (NPCC).

IOP Conf. Series: Materials Science and Engineering **395** (2018) 012009 doi:10.1088/1757-899X/395/1/012009

Acknowledgments
The authors would like to acknowledge the support for this work provided by PEKERTI Research grant, DRPM, Kemenristekdikti, 2017.

References

[1] ESDM Jawa Timur 2012 (www.pertambangan-jatim.or.id)
[2] Ramasamy V, Anand P and Suresh G 2017 *Int. J. of Mat. Sci.* **12** 499
[3] Varghese N, Harifaran M and Cherian A B 2015 *Wor. J. of Phar. Res* **4** 1302
[4] Othman R, Isa N and Othman A 2015 *Sai. Malay.* **44** 1561
[5] Ukrainczyk M, Kontrec J, Ivancic V B, Brecevic L and Kralj D 2006 *Pow. Tech.* **171** 192
[6] Teir S, Eloneva S and Zevenhoven R 2005 *Ener. Conv. & Manag.* **46** 2954
[7] Arifin Z, Pratapa S, Triwikantoro and Darminto 2013 *AIP Conf. Proc.* **1554** 90
[8] Arifin Z, Apriliani N F, Zainuri M and Darminto 2016 IOP *Conf. Series: Mat. Sci. And Eng.* **196** 1
[9] Arifin Z, Riyanto A, Lailiyah Q, Triwikantoro, Pratapa S, and Darminto 2015 *Trans. Ind. Ceram. Soc.* **74** 1
[10] Thriveni T, Ahn J W, Ramakrishna C, Ahn Y J and Han C 2016 *J. of the Kor. Phys. Soc.* **68** 1
[11] Harrison DJ 1993 *Quart. J. of Eng. Geo* **26** 293
[12] Mitchell CJ 2011 Brit. Geo. Surv. 48 (http://nora.nerc.ac.uk/15887/1/Reconnaissance_assesment_of_high-purity_limestone_plus_references.pdf)

Morphology Control of $MAPbI_3$ Perovskite Thin Film as An Active Layer of Solar Cells

Z R Pratiwi, L Nuraeni, A H Aimon, F Iskandar*
Department of Physics, Faculty of Mathematics and Natural Sciences, Institut Teknologi Bandung, Jalan Ganesa 10, Bandung 40132, Indonesia

*ferry@fi.itb.ac.id

Abstract. With one-step deposition, organic-inorganic perovskite solar cells with planar structure configuration can be produced. This technique has excellent potential for easy fabrication but it results in poor coverage of the film and thus affects the solar cells performance. In this research, morphology control was performed to improve the structure of the $MAPbI_3$ active layer by flowing hot air during one-step deposition. This allows the fabrication of simple, reproducible $MAPbI_3$ perovskite thin film with reduced pinhole formation. The hot airflow during the spin coating process homogeneously distributed grains of about 110 nm in size. A power conversion efficiency of 0.91 % was obtained with fill factor of 0.31 under a homemade sun simulator. The results indicate that hot airflow with reduced pinhole formation is a likely choice to control the morphology of thin film perovskite $MAPbI_3$ for solar cells.

Keywords: perovskite, solar cells, hot airflow, morphology control.

1. Introduction

Organic-inorganic perovskite materials have received much interest for scientific study and device application due to their unique optical, electronic, magnetic properties, and facile film processing [1-3]. Organic-inorganic perovskite materials are currently among the most competitive candidates for absorber materials in thin-film photovoltaic applications. In the past ten years remarkably high efficiencies of ~22% have been reported [4]. The reason for the rapid increase in power conversion efficiency (PCE) of such devices is that perovskite materials have an appropriate direct band gap, high absorption coefficient, and excellent carrier transport [1]. Therefore, one of the main challenges is the fabrication of high-quality films with controlled morphology, high surface coverage and minimum pinhole formation for high performance in perovskite devices.

One of the simplest techniques to fabricate solar cells is one-step deposition using spin coating on both the mesoscopic and planar structure. In previous studies, the planar structure of the perovskite active layer tends to produce higher pinhole density than the mesoscopic structure [5]. This affects low coverage, which causes low absorption in solar cell devices. However, the mesoscopic structure results in a less homogeneous film with the formation of islands during film growth on the TiO_2 [6]. Consequently, areas are uncovered by the perovskite layer [7]. A two-step sequential deposition method has been introduced but it results in a coarse morphology due to particles in the PbI_2 film. This technique is more suitable for solar cells with a mesoscopic structure [8].

A way to improve the reproducibility of perovskite $MAPbI_3$ is to control the morphological structure by increasing coverage and reducing pinhole formation. Fuzhi Huang reported the first gas-assisted

IOP Conf. Series: Materials Science and Engineering **395** (2018) 012010 doi:10.1088/1757-899X/395/1/012010

preparation of perovskite film in 2014 [13]. Perovskite thin film was grown assisted by gas (argon) during spin coating, obtaining a single crystal monolayer and thus increasing the solar cell efficiency with 16.5%. In 2017, Seulki et al. stated that hot airflow could induce the early steps of $MAPbI_{3-x}Cl_x$ crystal growth and yielded a PCE of 14.9% [10].

We have reported the morphological control of perovskite $MAPbI_3$ solar cells with a planar structure under high relative humidity (RH ±70%). The perovskite $MAPbI_3$ material is the optimum perovskite material for use as an active layer of solar cells because it has excellent stability [9]. Morphology control was done by applying hot airflow during the spin coating process. This was expected to improve the process of film growth by accelerating the evaporation process of $MAPbI_3$ solution. Thus, in general, attempts were made to improve the performance of solar cell devices with simple structures and processes.

2. Experiments

2.1Materials

MAI or CH_3NH_3I was synthesized by reacting 24 mL of methylamine (33 wt% in ethanol) and 10 mL of hydroiodic acid (57 wt% in air, Sigma Aldrich) and 100 mL ethanol absolute in a three-neck flask under nitrogen at 0 °C for 2 hours with stirring. After the reaction, the white precipitate of CH_3NH_3I was collected by homemade rotary evaporator at 60 °C. Lead (II) iodide (PbI_2) was purchased from Sigma Aldrich. SnO_2: Fluorine-doped (FTO), 7 Ohm/sqm, anhydrous N, N-dimethylformamide (DMF), titanium (IV) isopropoxide (TTIP), graphene was purchased from Sigma Aldrich.

2.2Solar Cell Fabrication

FTO glasses were cleaned in an ultrasonic bath with detergent, de-ionized water, acetone, 2-propanol, and ultrapure water for 5 minutes, respectively. The clean FTO was then dried by nitrogen gun. After being treated, a compact-TiO_2 layer was deposited onto the cleaned FTO substrates by spin coating a solution of titanium (I-V) isopropoxide (0.3 ml) in ethanol (5 ml) and hydrochloric acid (0.1 ml) with a speed of 3000 rpm and then sintered at 450 °C for 30 minutes. For the perovskite layer, a 0.88 M perovskite solution was prepared from PbI_2 and MAI (1:1-mole ratio) in DMF and spin-coated onto the FTO/TiO_2 substrate at 2000 rpm. After 5 s, hot airflow at 100 °C was blown over the surface of the perovskite films during the spin coating process. The flowing rate was varied at 1.5 m/s, 1.7 m/s, and 2 m/s. The distance of the substrate to the hot air nozzle was controlled at 2-3 cm. The perovskite films were then dried at 100 °C for 10 minutes on a hotplate. The final devices were completed by spreading 0.05 g of graphene contacts and putting FTO on top of the graphene contacts. The active area of the devices was 1 cm^2.

2.3Instruments and Characterization

X-ray diffraction (XRD) data were collected on Panalytical X'Pert Pro PW 3040 X-ray Powder Diffractometer. Scanning electron microscopy (SEM) was performed with field-emission electrons using an SEM Hitachi SU3500 instrument. Optical properties were measured with Spectra Suite and an Ocean Optic HR2000CG-UV-NIR instrument. The photovoltaic performance was characterized with an FTO encapsulation sandwich under illumination of a metal halide lamp in a homemade solar simulator, calibrated with silicon solar cell devices. The intensity was calculated using the lux to watts formula.

3. Results and Discussion

Morphology control of thin-film growth was carried out by providing a hot airflow at 100 °C during the process of one-spin coating using an additional heater kit. The effect of hot airflow during spin coating on the initial condition was analysed using an optical microscope with 1000x magnification. Figure 1 shows the initial conditions of the crystallization of $MAPbI_3$. The image was taken using an optical microscope at 10 seconds after the spin-coating deposition process.

IOP Conf. Series: Materials Science and Engineering **395** (2018) 012010 doi:10.1088/1757-899X/395/1/012010

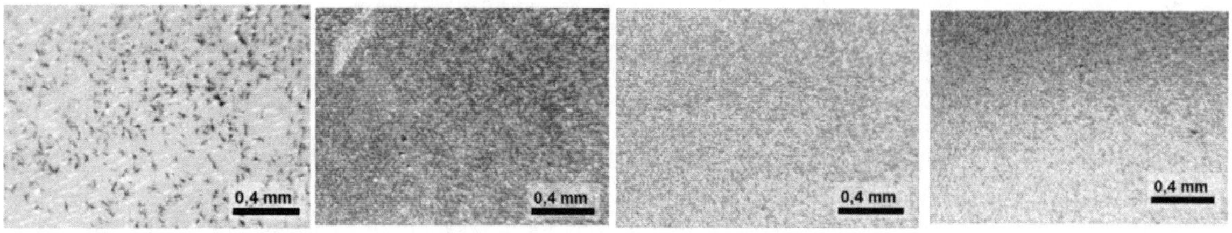

(a) (b) (c) (d)

Figure 1. Optical microscope images of initial stage of perovskite film crystallization: (a) without hot airflow (b) 1.5 m/s, (c) 1.7 m/s, and (d) 2 m/s of hot airflow rate.

The sample without hot airflow shows irregularly distributed needle-like shapes with low coverage. This is due to the DMF as a solvent not being entirely evaporated so that only a few initial particles were formed (Figure 1a). The sample with hot airflow at a rate of 1.5 m/s indicates uniformly distributed initial particles with high coverage and a lower size of initial particles. The presence of hot airflow during the spin coating caused the solvent to reach supersaturation. The samples with hot airflow at rates of 1.7 m/s and 2 m/s did not adequately show the morphology of the films. We conclude that the initial particle size became smaller and the solvent reached supersaturation faster. DMF has a boiling temperature of 153 °C, so the evaporation process on perovskite film will take a long time. In general, the process of precipitation of a solution onto a film involves two conditions: nucleation and crystal growth. Nucleation will be formed when the solution reaches supersaturation and the initial particles created will depend on the rate of nucleation and the degree of supersaturation. Thus, slow evaporation indicates a low degree of precursor supersaturation. This will cause crystal growth with a small number of particles and with aggregation on the surface as shown in Figure 1a. The flow of hot air during the deposition process is one of the approaches to control the state of precursor supersaturation as shown in Figure 1b. This is consistent with what has been reported in previous studies [10].

Furthermore, the samples were annealed at 100 °C for 10 minutes to finalize the crystal growth and were observed using a scanning electron microscope, as shown in Figure 2. The images show that the formed needle-like shapes indicate the growth of $MAPbI_3$ film. The films without hot airflow treatment had a large grain size (about 440 nm) and low coverage. The surface of FTO/TiO_2-c was not fully covered by the perovskite layer, which could decrease the device performance because the uncovered part or pinhole area would be a trap for electrons and cause recombination around the interface. Figure 2b shows the sample with hot airflow of 100 °C at a rate of 1.5 m/s resulting in needle-like shapes with a lower grain size. Besides, the resulting films show lower pinhole density.

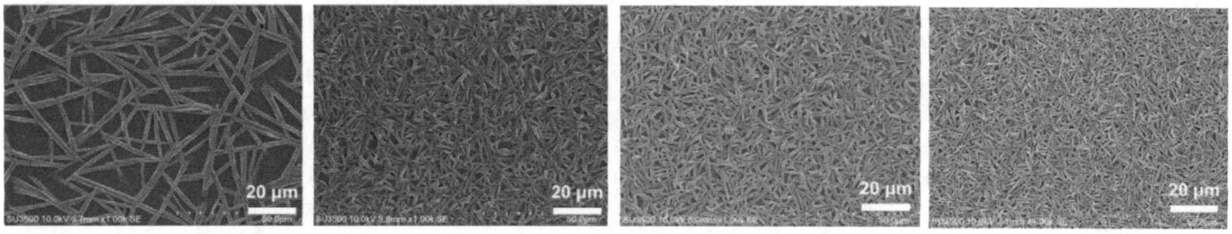

(a) (b) (c) (d)

Figure 2. SEM images of perovskite film (a) without hot airflow (b) 1.5 m/s, (c) 1.7 m/s, (d) 2 m/s of hot airflow rate.

The higher the flow rate given (Figure 2c and 2d), the denser the film surface, the lower the pinhole density and the smaller the crystal grain size, but the higher the coverage. High environmental humidity can cause a small grain size. The growth process of perovskite $MAPbI_3$ thin film is influenced by environmental humidity [11]. Low humidity can reduce grain boundaries because the grains aggregate more easily. However, the addition of hot airflow during the deposition process at high humidity leads to speeding up the crystallization process, resulting in a smaller crystal size but higher coverage. Hot airflow induces faster solidification of the film [12].

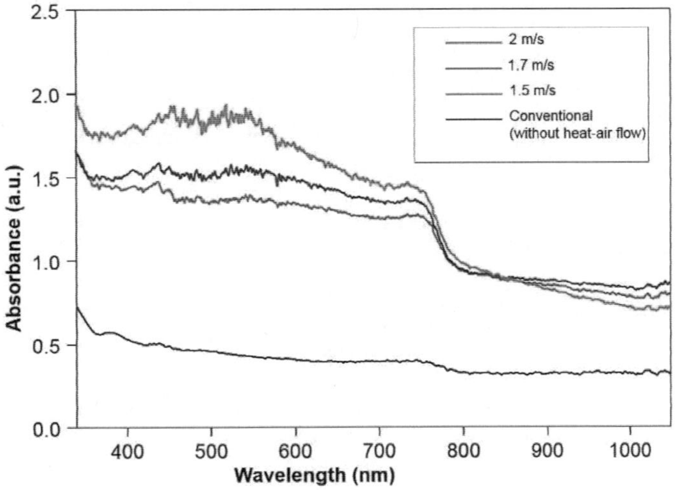

Figure 3. UV-Vis spectra of perovskite thin film.

Table 1. Results of Film Band Gaps Obtained from Absorbance Curve of UV-Vis-NIR Spectra

Hot airflow rate	Bandgap (eV)
0 m/s	1.44
1.5 m/s	1.51
1.7 m/s	1.52
2 m/s	1.51

The optical characteristics of the perovskite $MAPbI_3$ films were obtained by using UV-Vis spectrophotometric measurement. Figure 3 shows the absorbance curve and wavelength for each sample. It can be seen that the presence of hot airflow improved the absorbance of the perovskite films with absorbance onset at a wavelength of 750 nm, consistent with the report from Huang et al. [13]. The higher the hot airflow rate, the higher the film's ability to absorb light. This is related to the film morphology. The presence of hot airflow can improve film density as a function of the surface. Also, the band gap can be calculated from this curve, as shown in Table 1. Based on the Tauc-plot equation, bandgap calculations of 1.51 eV and 1.52 eV from the hot airflow treatment were obtained, while 1.44 eV was obtained from the film without hot airflow treatment. The presence of hot airflow caused a blue shift in the absorbance curve. As a consequence, there was a band gap difference between the samples. This is different from conventional semiconductor theory. The higher the temperature, the lower the resulting band gap. This is because of the occurrence of smaller valence band shifts due to the effect of temperature on $MAPbI_3$ films [14].

IOP Conf. Series: Materials Science and Engineering **395** (2018) 012010 doi:10.1088/1757-899X/395/1/012010

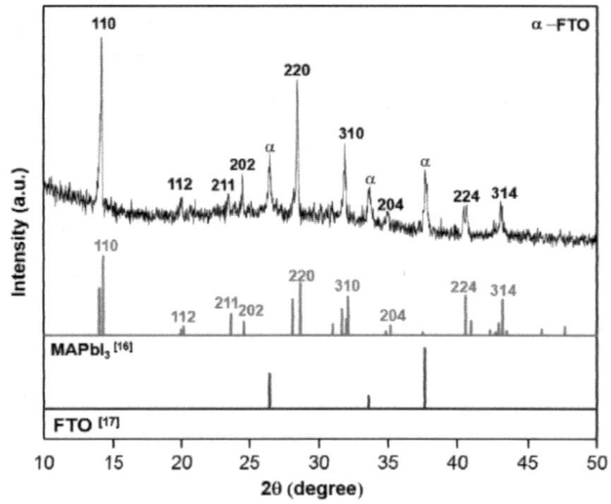

Figure 4. Crystal structure of $MAPbI_3$ film (diffraction peaks of FTO crystal marked by α).

XRD characterization of the films with a flow rate of 2 m/s (Figure 4) was conducted. Peaks at an angle of 2θ were obtained, corresponding to the reference, where the dominant peaks are (1 1 0), (2 2 0), and (3 1 0), respectively, at angles of 14.04°, 26.43°, and 31.81°. Based on the XRD results, it was found that the $MAPbI_3$ thin film had a tetragonal crystal structure [15].

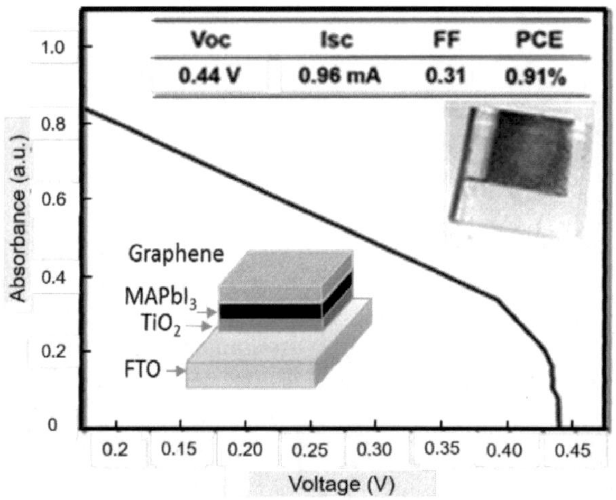

Figure 5. I-V characteristics of the device.

The optimum film morphology was obtained from a flow rate of 2 m/s. The I-V characteristics of the device fabricated were shown in Figure 5. The low value of PCE produced by the device could be caused by the recombination of electron-hole pairs before they reached the electrode. Furthermore, graphene powder as an electrode grown by spreading produces a less even layer, which could lead to electron traps.

4. Conclusions

In summary, hot airflow during spin coating deposition was developed to fabricate higher coverage of perovskite thin films as an active layer in solar cells devices under high humidity. This method facilitates faster evaporation of the solvent by increasing a degree of supersaturation. Consequently, more evenly distributed particles with low grain size were obtained and the morphology and optical characteristics of the film were improved.

IOP Conf. Series: Materials Science and Engineering **395** (2018) 012010 doi:10.1088/1757-899X/395/1/012010

5. References

[1] T. Ishihara 1995 *In Optical Properties Of Low-Dimensional Materials* (Singapore: World Scientific, ed. T. Ogawa, Y. Kanemitsu, 289-339)

[2] G. C. Papavassilious 1997 *Three- and Low-Dimensional Inorganic Semiconductors* (Prog. Solid State Chem, 25, 125-270)

[3] D. B. Mitzi 1999 *Synthesis, Structure, and Properties of Organic-Inorganic Perovskite and Related* (Prog. Inor. Chem, 48, 1-121)

[4] N. Park 2016 *Crystal Growth Engineering for High Efficiency Perovskite Solar Cells* (Cryst. Eng. Comm)

[5] G. E. Eperon, V. M. Burlakov, P. Docampo, A. Goriely, H. J. Snaith 2013 *Morphological Control for high Performance, Solution-Processed Planar Heterojunction Perovskite Solar Cells* (Adv. Func. Materials. 24, 1, 151-157)

[6] J. Burschka, N. Pellet, S. Moon, R. H.Baker, P. Gao, M. K. Nazeeruddin, M. Grätzel 2013 *Sequential Deposition as a Route to High-Performance Perovskite-Sensitized Solar Cells* (Nature 499, 316–319)

[7] J. H. Heo, S. H. Im, J. H. Noh, T. N. Mandal, C. S. Lim, J. A. Chang, Y. H. Lee, H. Kim, A. Sarkar, M. K. Nazeeruddin, M. Grätzel, S. I. Seok 2013 *Efficient Inorganic–Organic Hybrid Heterojunction Solar Cells Containing Perovskite Compound and Polymeric Hole Conductors* (Nature Photonics 7, 486–491)

[8] D. Liu, T. Kelly 2014 *Perovskite Solar Cells With a Planar Heterojunction Structure Prepared Using Room-Temperature Solution Processing Techniques* (Nature Photonic, 133-138)

[9] P. Gao, M. Gratzel, M. K. Nazeerudin 2014 *Organohalide Lead Perovskites for Photovoltaic Applications* (Energy Environmental Science, 7(8), 2448 – 2463)

[10] S. Song, M. T. Horantner, K. Choi, H. J. Snaith, T. Park. *J* 2017 *Inducing Swift Nucleation MorphologyControl for Efficient Planar Perovskite Solar Cells by Hot Air Quenching* (Mater. Chem. A. 5, 3812-3818

[11] J. You, Y. M. Yang, Z. Hong, T. Song, L. Meng, Y. Liu, C. Jiang, H. Zhou, W. Chang, G. Li, Y. Yang 2014 *Moisture Assited Perovskite Film Growth For High Performance Solar Cells* (Appl. Phys. Lett. 105, 183902)

[12] S. Bae, S. J. Han, T. J. Shin, W. H. Jo 2013 *Two different Mechanisms of $CH_3NH_3PbI_3$ Film Formation in One-step Deposition and its effect on Photovoltaic Properties of OPV-type Perovskite Soalr Cells* (J. Mater. Chem. A., 3, 23964-23972)

[13] F. Huang, Y. Dkhissi, W. Huang, M. Xiao, I. Benesperi, S. Rubanov, Y. Zhu, X. Lin, L. Jiang, Y. Zhou, A. Gray-Weale, J. Etheridge, C. R. McNeill, R. A. Caruso, U. Bach, L. Spiccia, Y. Cheng 2014 *Gas-assisted Preparation of Lead Iodide Perovskite Films Consisting of Monolayer of Single Crystalline Grains for High Efficiency Planar Solar Cells* (Nano Energy 10, 10-18)

[14] A. Slonopas, B. J. Foley, J. J. Choi, M. C. Gupta 2016 *Charge Transport in Bulk CH3NH3PbI$_3$ perovskite* (Journal of Appl. Phys. 119, 074101)

[15] T. Oku 2015 *Crystal Structures of $CH_3NH_3PbI_3$ and Related Perovskite Compounds Used for Solar Cells* (Energy Engineering ISBN 978-953-51-2184-8)

IOP Conf. Series: Materials Science and Engineering **395** (2018) 012011 doi:10.1088/1757-899X/395/1/012011

Synthesis of Nanocrystalline $Ba_{0,5}Sr_{0,5}FeO_{3-\delta}$ by Sol-gel Self-Combustion Method

F Fitriana, D Z Anjarwati, P S N Baity, and S Suasmoro*

Dept. of Physics, Institut Teknologi Sepuluh Nopember Surabaya 60111,Indonesia

* Corresponding author: suasm@its.ac.id

Abstract. Fuel cell is a one of device to convert chemical into proper electric energy, has some advantages including high efficiency, long term stability, fuel flexibility, environment friendly. To develop SOFC operates at intermediate temperature (500-700ºC), it is necessary to figure out a cathode material capable of overcoming the low electrochemical reduction activity in oxygen at intermediate temperatures. $Ba_{0,5}Sr_{0,5}FeO_{3-\delta}$ (BSF) has mixed ionic and electronic conductivity properties (MIEC). A sol-gel self-combustion method was used to synthesize BSF due to easier preparation and produces excellent result (high homogeneity and high reactivity) powder. Two conditions were set up in different pH, namely 0-1 and 6-7. The self-combustion reaction occurred on heating up to 220 ºC and produced yield ashes. To acquire the calcination condition, the temperature was carried out at 500, 700, 900, 1000, and 1100 °C for 6 hours. In this study, the phase formation, structure, and crystallite size of BSF was characterized by make use of X-ray diffraction (XRD) data. The results show that BSF has cubic perovskite structure with space group *pm3m* (221), crystallite size 52 and 109 nm for pH 0-1 and 6-7 respectively.

Keywords: MIEC, perovskite, sol-gel self-combustion method.

1. Introduction

The perovskite oxide (ABO_3) is promising structure for MIEC material, due to BO_6 octahedra playing a role in ionic and electronic conductivity[1]. Cobalt-based perovskite oxides such as $La_{0,6}Sr_{0,4}Co_{0,2}Fe_{0,8}O_{3-\delta}$ and $Ba_{0,5}Sr_{0,5}Co_{0,8}Fe_{0,2}O_{3-\delta}$ have been widely used and reported to have high power density[2]. However, cobalt-based cathodes have some disadvantages: low chemical stability in reduced environments, large volume change due to cobalt ion spin transitions at the time of reduction, and reacts with zirconia-based electrolytes[3]. Therefore, from several options, the Fe (iron) -based cathode which has MIEC characteristic attracted much attention because of its catalytic activity similar to that of cobalt-grade perovskites, thermal stability and higher structure, low cost, and abundant presence[4,5]. Fe-based perovskite showed better stability than perovskite-based cobalt[5]. However, the Fe transition metal has several variables of oxidation and spin states, potentially affecting high catalytic activity and high resistance at the cathode[6]. $BaFeO_{3-\delta}$ has superior Oxygen Reaction Reduction (ORR) activity due to a high concentration of oxygen vacancies in their lattice structures[7]. In addition, Yin also reported that Sr doping provides the highest ratio between absorbed and surface oxygen[8] and a large ionic radius and a low valence state of Ba^{2+} are excellent in generating oxygen vacancies for oxygen transport[7].

Several methods are often used to synthesize SOFC electrodes that are solid state reaction[9], sol-gel[10,11], and self-combustion[7,12–16]. It is well known that solid state reaction process need a high temperature of calcination. The sol-gel method are complicated to do and the cost is high[14]. Combustion was chosen to synthesize perovskite $BaFeO_{3-\delta}$ due to easier in precursor preparation and produces excellent product with high homogeneity and high reactivity powder[12]. In addition, the combustion method produces a powdered product of nanocrystalline size[13].

In this study, we develop perovskite material based on $Ba_{0,5}Sr_{0,5}FeO_{3-\delta}$ as the basic material to develop new cathode material. We focus on the synthesis of $Ba_{0,5}Sr_{0,5}FeO_{3-\delta}$ powder by sol-gel self-combustion method. In this paper, we report the synthesis process, the phase formation, and crystal size which determined by X-ray diffraction.

2. Experimental

$Ba_{0,5}Sr_{0,5}FeO_{3-\delta}$ (BSF) powders were synthesized using sol-gel self-combustion method. Stoichiometric amount of $Ba(NO_3)_2$, $Sr(NO_3)_2$, and $Fe(NO_3)_3.9H_2O$ were dissolved in deionized water at 30°C. And $C_6H_8O_7.2H_2O$ solution were added under stirring to form an aqueous solution. The pH value was varied at 0-1 and 6-7 by adding ammonium solution. The mixed solution then heated at 100 °C and increased gradually at 150, 200, and 220 °C to obtain a dark solid precursor. The BSF powder was milled, and followed by calcination at 500, 700, 900, 1000, and 1100 °C for 6h in air to generate the desired perovskite oxide. The phase formation, structure and crystallite structure were determined by X-ray diffraction (XRD).

3. Results and Discussion

It was observed that all gels combusted into black powder of perovskite at 220°C. In sol-gel self-combustion method, metal ions react with COO^- in the solution to form metal-citrate complex. One of the factor that affect sol formation and its stability is pH value. Addition of ammonium solution for pH adjustment affect the dissociation of citric acid and metal complex formation as well. But, addition of ammonium solution should not precipitate and can destroy the balance of all reaction. Sol-gel self-combustion method of $Ba_{0,5}Sr_{0,5}FeO_{3-\delta}$ is based on the principle of oxidation and reduction, expressed by following reaction:

$$0.5Ba(NO_3)_2 + 0.5Sr(NO_3)_2 + Fe(NO_3)_3 + 5/3C_6H_8O_7 + 3O_2 \rightarrow Ba_{0,5}Sr_{0,5}FeO_{3-\delta} + 20/3\ H_2O + 10CO_2 + 5/2\ N_2$$

Table 1. Stability of sol and combustion degree with varied pH value.

	Sample Number	
	1	2
Ph	0-1	6-7
State of sol	Clear	Clear
Color of gel	Red-brown	Dark green
Combustion	Happened rapidly at 220 °C, some visible flame	Happened fast at 220 °C, no visible flame
Color of product	Brown and black powder	Brown and black powder

Figure 1 and 2 show the XRD pattern of $Ba_{0,5}Sr_{0,5}FeO_{3-\delta}$ at different temperature of calcination and varied pH value. The XRD pattern with pH 0-1 shows in Figure 1, which showed that the powder has formed crystallite after combustion with orthorhombic structure. The phase formation and composition of the powder shows in table 2.

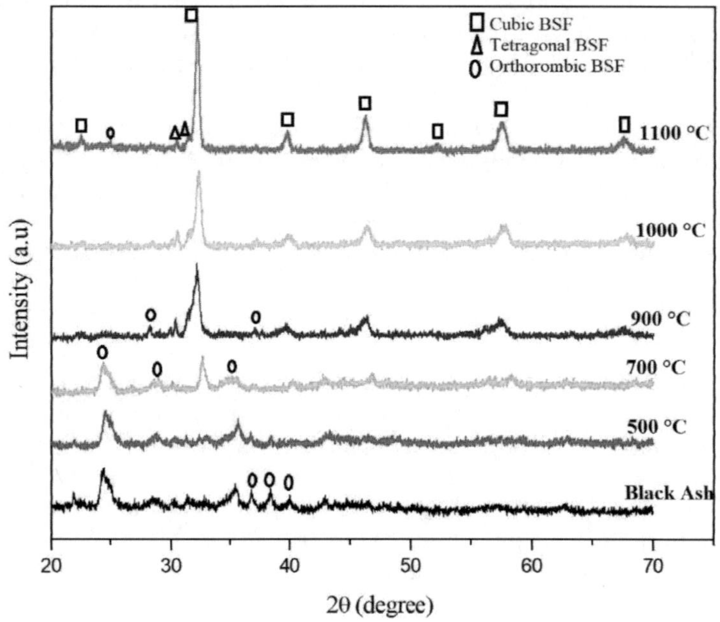

Figure 1. XRD Pattern of $Ba_{0,5}Sr_{0,5}FeO_{3-\delta}$ pH: 0-1 after 6 hours calcination at different temperature.

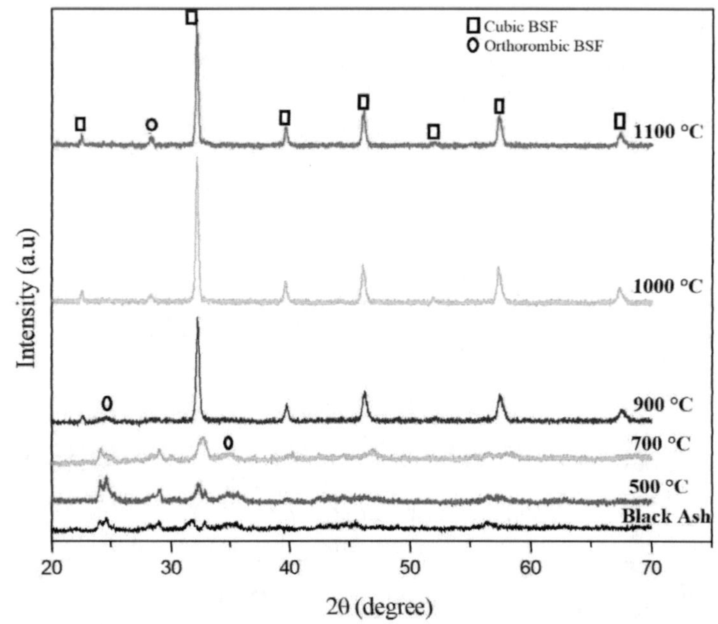

Figure 2. XRD Pattern of $Ba_{0,5}Sr_{0,5}FeO_{3-\delta}$ pH: 6-7 after 6 hours calcination at different temperature.

The phase transformation from orthorhombic to cubic occur at 900°C based on XRD pattern for both pH value. XRD data shows the crystallization process at BSF with ammonium solution addition (pH: 6-7) faster than without ammonium solution (pH: 0-1). It is related to the influence of ammonium solution addition than can affect the dissociation of citric acid and metal complex formation. The cubic phase reaches 98,99 % at 900 °C for 6h when pH 7. This result is better than result reported by Toprak et.al [17] using oxalate co-precipitation. Another method also has been reported using solid state reaction [18,19] but the calcined and sintered temperature were high. The result of Rietveld refinement method shows in Figure 3.

Table 2. Relative weight fraction of phases percent of $Ba_{0,5}Sr_{0,5}FeO_{3-\delta}$ (BSF) at different calcination temperature by Rietveld refinement.

Temperature	Relative weight fractions (%)		
		pH = 0-1	
	Cubic BSF (%)	Orthorhombic BSF (%)	Tetragonal BSF (%)
220 °C (Black ash)	-	100	-
500 °C	-	100	-
700°C	-	80.98	19.02
900°C	95.08	4.69	0.24
1000°C	96.96	1.95	1.09
1100°C	98.13	1.26	0.61
		pH = 6-7	
220 °C (Black ash)	-	100	-
500 °C	-	100	-
700°C	-	80.98	19.02
900°C	98.99	1.01	-
1000°C	96.73	3.27	-
1100°C	98.01	1.99	-

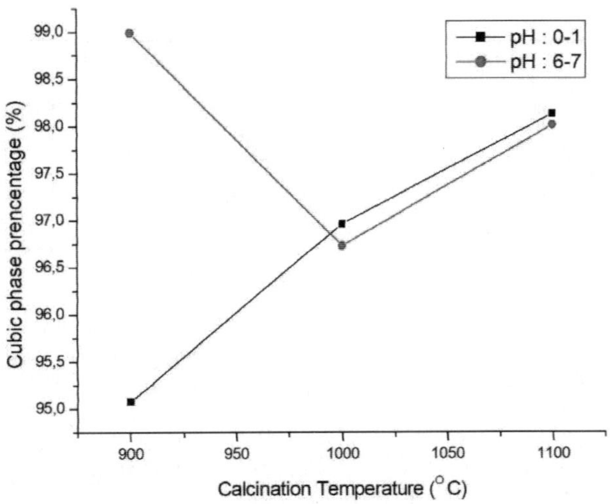

Figure 3. Cubic phase percentage related to the temperature of calcination for both pH value.

To analyze the change of lattice parameter in cubic, we used Rietveld refinement method. The result is shown on table 3. Based on Rietveld refinement result, the cubic phase (space group: *pm3m* (221)), the average of lattice parameter is 3,91-3,93 (Å).

Table 3. Lattice Parameter of $Ba_{0,5}Sr_{0,5}FeO_{3-\delta}$ using Rietveld Refinement.

Calcination Temperature	Lattice Parameter (Å)		
		pH = 0-1	
	a (Å)	V (Å)	FoM
900 °C	3.9305 (3)	60.7222	Rbragg: 2,44; 5,77; 6,99 Rp: 7,71; Rwp: 9,78 Rexp: 7,53 GoF: 1,69
1000 °C	3.9143 (1)	59.9754	Rbrag: 1,96; 8,28; 9,06 Rp: 8,01; Rwp: 10,40 Rexp: 7,97; GoF: 1,70
1100 °C	3.9277 (7)	60.3622	Rbragg: 1,62; 4,51; 3,70 Rp: 7,03; Rwp: 8,93 Rexp: 7,41; GoF: 1,45
		pH = 6-7	
	a (Å)	V (Å)	FoM
900 °C	3.9224 (7)	60.3481	Rbragg: 2,71; 1,99 Rp: 6,40; Rwp: 8,12 Rexp: 7,53; GoF: 1,69
1000 °C	3.9304 (6)	60.7207	Rbragg: 1,56; 5,10 Rp: 6,55; Rwp: 8,19 Rexp: 7,08; GoF: 1,33
1100 °C	3.9266 (4)	60.5423	Rbragg: 1,64; 8,27 Rp: 7,08; Rwp: 9,22 Rexp: 7,81; GoF : 1,39

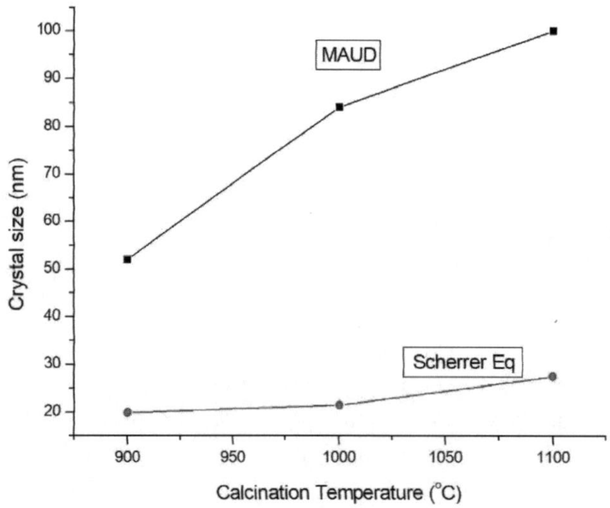

Figure 4. Crystal size related to the temperature of calcination at pH = 0-1.

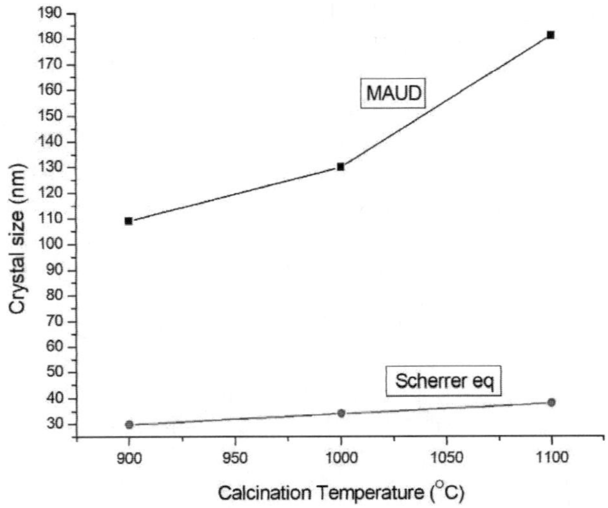

Figure 5. Crystal size related to the temperature of calcination at pH = 6-7.

The crystallite size was calculated by MAUD software and Scherrer equation. The result show that the crystal size is growing up by increasing temperature. The result is more accurate by using MAUD software than Scherrer equation.

4. Conclusion

$Ba_{0,5}Sr_{0,5}FeO_{3-\delta}$ cubic phase was obtained through sol-gel self-combustion method with calcination process at 900 °C for 6h for both pH value which varied. The smaller crystal size, 50-55 nm, achieve when the pH value is 0-1.

5. Reference

[1] Kim D-Y, Miyoshi S, Tsuchiya T and Yamaguchi S 2014 Electronic Defect Formation in Fe-Doped $BaZrO_3$ Studied by X-Ray Absorption Spectroscopy *Chem. Mater.* **26** 927–34

[2] Shan D, Gong Z, Wu Y, Miao L, Dong K and Liu W 2017 A novel $BaCe_{0.5}Fe_{0.3}Bi_{0.2}O_{3-\delta}$ perovskite-type cathode for proton-conducting solid oxide fuel cells *Ceram. Int.* **43** 3660–3

[3] Wang J, Lam K Y, Saccoccio M, Gao Y, Chen D and Ciucci F 2016 Ca and In co-doped $BaFeO_{3-\delta}$ as a cobalt-free cathode material for intermediate-temperature solid oxide fuel cells *J. Power Sources* **324** 224–32

[4] Zhang C and Zhao H 2012 A novel cobalt-free cathode material for proton-conducting solid oxide fuel cells *J. Mater. Chem.* **22** 18387–94

[5] Wang J, Saccoccio M, Chen D, Gao Y, Chen C and Ciucci F 2015 The effect of A-site and B-site substitution on $BaFeO_{3-\delta}$: An investigation as a cathode material for intermediate-temperature solid oxide fuel cells *J. Power Sources* **297** 511–8

[6] Yang G, Shen J, Chen Y, Tadé M O and Shao Z 2015 Cobalt-free $Ba_{0.5}Sr_{0.5}Fe_{0.8}Cu_{0.1}Ti_{0.1}O_{3-\delta}$ as a bi-functional electrode material for solid oxide fuel cells *J. Power Sources* **298** 184–92

[7] Dong F, Chen Y, Chen D and Shao Z 2014 Surprisingly High Activity for Oxygen Reduction Reaction of Selected Oxides Lacking Long Oxygen-Ion Diffusion Paths at Intermediate Temperatures: A Case Study of Cobalt-Free $BaFeO_{3-\delta}$ *ACS Appl. Mater. Interfaces* **6** 11180–9

[8] Porras-Vazquez J M, Smith R I and Slater P R 2014 Investigation into the effect of Si doping on the cell symmetry and performance of $Sr_{1-y}Ca_yFeO_{3-\delta}$ SOFC cathode materials *J. Solid State Chem.* **213** 132–7

[9] Chandratreya S S, Fulrath R M and Pask J A 1981 Reaction Mechanisms in the Formation of PZT Solid Solutions *J. Am. Ceram. Soc.* **64** 422–5

[10] Budd K D, Dey S K and Payne D A 1985 Sol-Gel Processing of $Pbtio_3$, $Pbzro_3$, PZT, AND Plzt

Thin Films, *British Ceramic Proceedings*, **36** 107-121
[11] Yi G and Sayer M 1996 An acetic acid/water based sol-gel PZT process I: Modification of Zr and Ti alkoxides with acetic acid *J. Sol-Gel Sci. Technol.* **6** 65–74
[12] Schäfer J, Sigmund W, Roy S and Aldinger F 1997 Low temperature synthesis of ultrafine $Pb(Zr, Ti)O_3$ powder by sol-gel combustion *J. Mater. Res.* **12** 2518–21
[13] Li Y, Xue L, Fan L and Yan Y 2009 The effect of citric acid to metal nitrates molar ratio on sol–gel combustion synthesis of nanocrystalline LaMnO3 powders *J. Alloys Compd.* **478** 493–7
[14] Guo R S, Wei Q T, Li H L and Wang F H 2006 Synthesis and properties of $La_{0.7}Sr_{0.3}MnO_3$ cathode by gel combustion *Mater. Lett.* **2** 261–5
[15] Xiao S H, Jiang W F, Li L Y and Li X J 2007 Low-temperature auto-combustion synthesis and magnetic properties of cobalt ferrite nanopowder *Mater. Chem. Phys.* **106** 82–7
[16] Gómez-Cuaspud J A and Vera-López E 2017 Synthesis and characterization of $La_{0.8}Sr_{0.2}Ni_{(1-x)}Cr_xO_3$ (x = 0.0, 0.2, 0.4, 0.6, 0.8, 1.0) system by the combustion method *Boletín la Soc. Española Cerámica y Vidr* **56**(6) 273-282
[17] Toprak M S, Darab M, Syvertsen G E and Muhammed M 2010 Synthesis of nanostructured BSCF by oxalate co-precipitation - As potential cathode material for solid oxide fuels cells *Int. J. Hydrogen Energy* **35** 9448–54
[18] Dong, F, Chen, Y, Ran, R, Chen, D, Tadé, M.O, Liu, S, Shao, Z 2013 $BaNb_{0.05}Fe_{0.95}O_{3-\delta}$ as a new oxygen reduction electrocatalyst for intermediate temperature solid oxide fuel cells *J. Mater. Chem. A* **1**(34) 9781–9791
[19] Zhu M, Cai Z, Xia T, Li Q, Huo L and Zhao H 2016 Cobalt-free perovskite BaFe0.85Cu0.15O3-δ cathode material for intermediate-temperature solid oxide fuel cells *Int. J. Hydrogen Energy* **41**(8) 4784-4791

Prospect and Challenges of Tight Ultrafiltration Membrane in Drinking Water Treatment

P T P Aryanti[1], A N Hakim[2], S Widodo[2,3], I N Widiasa[4], I G Wenten[2,*]

[1]Chemical Eng. Dept., Jenderal Achmad Yani University, Jl. Terusan Jend. Sudirman PO BOX 148, Cimahi, Indonesia.
[2]Department of Chemical Engineering, Institut Teknologi Bandung, Jl. Ganesha 10, Bandung 40132, Indonesia.
[3]PPPTMGB LEMIGAS, Balitbang KESDM, Jl. Ciledug Raya, Kav. 109, Cipulir, Kebayoran Lama, Jakarta, 12230, Indonesia.
[4]Chemical Engineering Department, Diponegoro University, Jl. Prof Sudarto-Tembalang, Semarang 50239, Indonesia.

*Corresponding author: igw@che.itb.ac.id

Abstract. Ultrafiltration (UF) membrane for drinking water treatment has been consistently modified to improve the selectivity towards trace organic compounds (TrOCs) and maintain the flux stability for long-term application. Currently, the tight UF membrane has been developed to answer the challenges of conventional UF membranes. With small pore size (300 - 5000 Da), the tight UF showed higher TrOCs rejection and lower irreversible fouling than the conventional UF membranes. Two methods have been proposed to fabricate the low molecular weight cut off (low-MWCO) UF membranes, i.e. blending membrane solution with fillers (nanoparticles or volatile solvent) and modifying membrane surface layer by chemical or plasma grafting method. Low water flux and high operating pressure are some challenges to be solved during the application of the tight UF membrane. Hence, the development of tight UF membrane with low-MWCO pore size has focused on improving flux while maintaining high selectivity. In this paper, the development and prospect of tight UF membrane will be reviewed. Challenges during the application of low-MWCO UF membrane for water treatment and methods to solve the challenges will be discussed.

Keywords: drinking water, high selectivity, low-MWCO, trace organic, ultrafiltration.

1. Introduction

In drinking water treatment, pressure driven membrane technologies achieved extraordinary levels of process intensification, substantially decreasing the equipment-size/production-capacity ratio, energy consumption, and waste production, which result in lower cost and sustainable technical solutions [1- 3]. In spite the advantages, the application of pressure driven membrane in water treatment is limited by fouling phenomenon, which reduces the production efficiency [4-8]. Details of the theoretical background of the pressure-driven membrane processes including strategies to prevent and mitigate membrane fouling have been reported in several literature [9-16]. Reverse osmosis (RO) is the most used membranes in surface water treatment and desalination to produce fresh (pure) water with low content of soluble organics and ions [17-19]. However, the demineralized water by RO has not been recommended as drinking water for long-term consumption related to the World Health Organization

(WHO) regulation [20]. The possible health consequences of low mineral content water consumption have been discussed in the literature [21, 22]. Hence, remineralization of NF and RO water product to reduce the water corrosivity is required before the water is distributed to the customer [23]. The standard for drinking water quality in terms of total dissolved solids (TDS) and mineral content has been summarized in the WHO guideline [23-25]. UF is an attractive alternative in water and drinking water treatment, which exhibits some advantages compared to NF and RO, i.e. lower investment cost, higher permeate flux, and lower energy consumption. From an economic point of view, the installation of UF membranes in water treatment plant may significantly reduce the operating cost by eliminating coagulation, flocculation, and sedimentation processes.

The UF membranes are generally characterized by their pore size between 0,01 μm (1 nm) – 0,1 μm (100 nm) or its molecular weight cut-off (MWCO) between 10,000 - 100,000 Dalton (Da) [25]. Due to its small pore size, the UF membrane provides an effective barrier for suspended particles, colloids, and pathogens while keeping the mineral content in water intake. Up to this time, the modification of UF membranes is mostly focused on improving the hydrophilicity, to reduce energy consumption and enhance productivity [26-30] and enhance an anti-fouling property of the membrane to maintain the membrane flux for long-term use [31]. Numerous review and research studies related to the UF membranes for water and drinking treatment have been published, as shown in Figure 1 [32- 38]. Simultaneously, development of UF membranes with innovative modules is also growing significantly to fulfill the requirement of fresh or drinkable water [39-42]. The development is not only for industrial and urban areas, but also for a rural and remote area that have inadequate sanitation, poor hygiene habits, and lack of adequate quantities of water for hygiene [43-46].

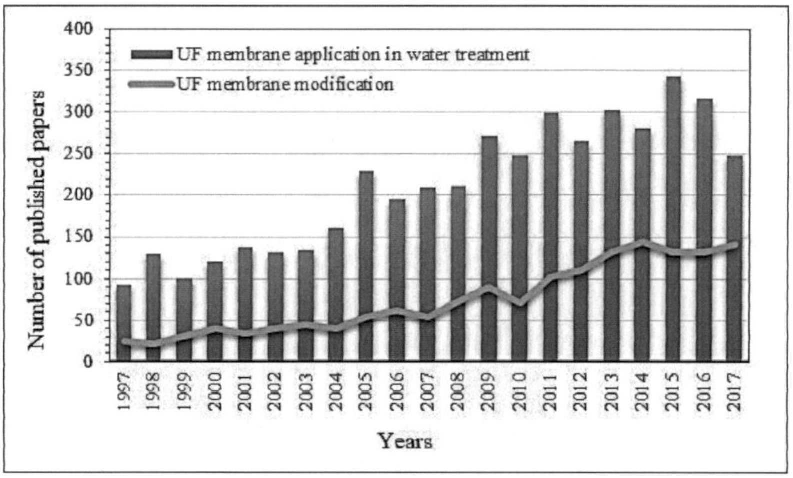

Figure 1. Number of publication related to UF membrane application in water treatment indexed by Scopus (TITLE+ABS+KEY: ultrafiltration membrane and water treatment).

Beside intrinsic fouling phenomenon, another limitation of UF membrane in water treatment, particularly for drinking water, is low rejection against soluble contaminants (up to 50%), such as trace organic compounds (TrOCs) with an approximate molecular weight below 1000 Da. The TrOCs, such as endocrine disrupting compounds (EDCs) and pharmaceuticals or personal care products (PPCPs), have been identified as harmful pollutants in drinking water treatment since they are suspected to be carcinogenic [47]. Tight UF membranes have been developed to overcome the limitation in conventional UF membranes. This membrane has a tight structure in the skin layer, with pore sizes between 300 and 5000 Da [48-52]. Due to its tight skin layer, the UF membrane has a high rejection against soluble compounds in water, such as humic substances and dyes [53, 54]. Up to this time, there exists little literature that discusses the tight UF membrane and its application for water treatment. In this paper, the low-MWCO UF membrane will be reviewed comprehensively including the state of the art and preparation method. In addition, challenges during the application of low- MWCO UF membrane for water treatment and methods to solve the challenges will be discussed.

2. The Application of Tight UF Membrane in Water Treatment

Several studies have been conducted to investigate the performance of tight UF membranes when it was applied in water treatment. One of the studies was conducted by Kramer et al. [54]. A tight UF membrane, with a pore size of 3000 Da, was used as pre-treatment of RO membrane to produce demineralized water from raw municipal sewage. It was found that the tight UF has an ability to reject chemical oxygen demand (COD) by 80%. Lin et al. [55] investigated the selectivity of a tight UF membrane, with a pore size of 4700 Da, to separate mixtures of dyes and Na_2SO_4. The tight UF membrane showed high rejection of dyes up to 98%. Similar research was carried out by Ma et al. [56], where the tight UF was used to separate dyes and mixed salts (both $NaCl/Na_2SO_4$). High rejection of dyes was also achieved (>98%). The pure water permeability of tight-UF membrane was found at least 6 times that of NF membrane. These results showed that the tight UF membrane can be used as an alternative to NF membranes.

Efligenir et al. 57] used two types of UF tight membrane (1000 Da), i.e. flat-sheet polyamide and TiO_2 tubular ceramic membrane, to remove metal salts from industrial discharge water, and then compared the performance of two membranes. They found that the TiO_2 ceramic membrane showed higher rejection of some metal ions, i.e. above 80% at operating pressure of 8 bar, while the polyamide membrane rejection was only 40%. It suggested that the negative surface charge of TiO_2 ceramic membrane played an important role to the metal ion rejection. Shang et al. [58] also used TiO_2 tight ceramic UF membrane (3000 Da) as pre-treatment of RO membrane to remove phosphate in water. They investigated the influence of membrane charge on the separation efficiency of phosphate by regulating the feed water condition, such as pH and salt concentration. The results pointed out that the ceramic membrane was more negatively charged in higher pH condition, which resulted in higher rejection of phosphate. Almost 87% rejection of phosphate can be achieved at pH 8.5. Meanwhile, the presence of Na^+ and Ca^{2+} ions in solution decreased the zeta potential of the membrane and led to the increase of natural organic matter (NOM) or effluent organic matter (EFOM) on the membrane surface [59]. The phosphate rejection was found to be correlated to the zeta potential of the organics in the feed water. The application of tight UF in water treatment is summarized in Table 1. Due to its high selectivity at low energy consumption as well as low irreversible fouling, certainly, the tight UF membrane can be considered as a future drinking water treatment technology.

Table 1. The applications of tight UF membranes in water treatment

Membrane material	UF pore size	Application	Operating condition	Results	Ref.
Polysulfone (PSf)	-	Humic substance removal in peat water	Operating pressure: 1 bar	Humic rejection: >80%; Flux: 126 LMH; Low irreversible fouling	[52]
Ceramic membrane	3000 Da	Sewage water treatment	Operating temp.: 20°C Operating pressure: 8 bar	Organic matter rejection: 81%	[54]
Polyethersulfone (PES)	4700 Da	N_2SO_4/dye separation	Operating temp.: 25 °C cross-flow rate : 60 L/h process mode: filtration-diafiltration Operating pressure: 4 bar	Na_2SO_4 rejection: >98%	[55]
Ceramic Membrane (TiO_2/ZrO_2 active layer)	8800 Da	Separation of dyes and mixed salts (NaCl/Na_2SO_4)	Flow rate: 3 m/s at 25°C Operating pressure: 2 bar	Na_2SO_4 rejection: >98% NaCl rejection < 10%	[56]
Ceramic membrane (TiO_2)	3000 Da	Separation of phosphate	pH of solution: 6 – 8.5	87% of phosphate rejection at pH 8.5	[58]
Regenerated cellulose acetate (RC)	5000 Da	2,4,6-trichloroanisole (TCA) removal	Cross flow rate: 0,7 ml/min Operating pressure: 4 bar	TCA rejection: > 80%	[60]
Charge ceramic membrane	3000 Da	Phosphate removal during WWTP secondary effluent	pH 7.0 ± 0.5 cross-flow velocity: 1 m/s flux: 30 Lm-2h-1	Phosphate rejection: 62%	[59]
Ag-doped TiO_2/Ti composite membrane	3000 Da	Dye removal (chlorazol fast pink, chlorazol black, and titan yellow)	Operating pressure: 1-7 bar Operating temp.: 25°C	Dye rejection: 99%	[61]
Modified poly(vinyl butyral) (PVB) hollow fiber membranes with carbon nanotubes (CNTs)	-	Humic substance removal	-	Humic substance rejection: >90%	[62]

3. Tight UF Membrane Preparation

A few studies have been focused on the tight UF membrane preparation techniques, including the dispersing fillers (polymeric or inorganic) in membrane structure 63,64] and membrane surface modifications by chemical grafting or pore-filling method [65-67]. Among these methods, blending membrane materials with hydrophilic and anti-fouling nanoparticles has attracted much attention in tight UF membrane manufacturing along with refusing fouling from the membrane surface [68-70]. Another method has been proposed by filling the membrane pore with a high cross-link polymer through in-situ polymerization [71,72] or with nanoparticles by vacuum filtration [73]. In the following subchapter, the preparation methods of tight UF membrane will be discussed.

3.1. Tight UF polymeric membrane preparation by dispersing fillers in membrane structure

There are two methods to disperse the fillers in membrane structure, namely ex-situ and in-situ method. In ex-situ method, the fillers are dispersed by blending into the membrane solution, while in in-situ method, the fillers are directly generated in membrane solution through a chemical reaction. Blending polymer solution with hydrophilic or anti-fouling additives are widely used in the manufacture and modification of membranes used in UF membrane modification due to its simple and inexpensive method. The frequent additives used in membrane modification are hydrophilic polymer (e.g. polyvinyl pyrrolidone or PVP and polyethylene glycol or PEG), organic compounds (e.g. glycerol, alcohol), inorganic salts (e.g. $LiNO_3$), inorganic compound (e.g. zeolite, TiO_2, Ag), water, and recently modified polymeric additive (e.g. sulfonated polyether ether ketone or SPEEK and PVP- co-styrene). In tight UF membrane preparation, some of these additives have been used and will be discussed in this sub-chapter.

Sadrzadeh et al. [74] manufactured UF membrane with a pore size of 2000 Da by blending 15%wt of PES and 5% wt of PVP (1300 Da). Further addition of PVP to 10%wt reduced the membrane pore size to 1000 Da and changed the membrane morphology from finger type to sponge type. Aryanti et al. [75] prepared a low-MWCO UF membrane, which had a high rejection to humic substances (up to 80%). They mixed PSf solution with acetone to obtain tight surface layer and PEG400 to increase membrane hydrophilicity. The rapid loss of acetone from the membrane solution lead to a higher polymer concentration in the membrane skin layer and produced tight pore structures [76]. As shown in Figure 2, the presence of acetone in membrane solution improved the humic substance rejection [77]. Furthermore, a tight skin layer of the membrane weakens the interaction between organic matters and the membrane surface, which is easily cleaned by flushing method and contribute to higher flux recovery ratio (FRR) [52].

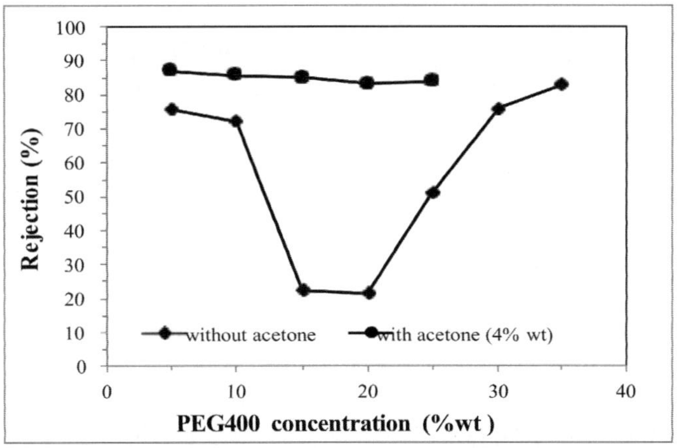

Figure 2. Effect of acetone concentration to humic substances removal 77].

Recently, blending polymer solution with inorganic nanoparticles becomes a choice of numerous researchers to produce a smaller membrane pore size with a good pore distribution along the thickness of UF membrane. TiO_2 has been widely used as an additive due to its properties and good compatibility with organic solvent during the preparation of UF membrane. It has been reported that composite UF membrane with TiO_2 is able to provide high rejection of low MWCO organic matter and good anti-fouling performance. Hamid et al. 78] found that polysulfone (PSf)/TiO_2 composite membrane was able to perform up to 90% rejection of low molecular humic acid. However, the pore size of the resulted membrane was beyond the pore range of Tight UF. Leo et al. [79] found that almost 80% permeate flux can be maintained during filtration process of humic solution, due to the presence of zeolite in membrane structure. Another hydrophilic nanoparticles, such as ZrO_2 [80], SiO_2 [81], zeolite [82], ZnO [83, 84], and Ag [85, 86], are also considered as an alternative modifier to enhance the performances of UF membranes [79]. By increasing the nanoparticle additive concentration, the smaller finger-like pore is formed due to the exchange rate between solvent and nonsolvent has been delayed thus resulted in a tight pore structure and higher rejection of foulants. Kotte et al. [72] prepared a tight PVDF UF membrane by in-situ generation of crosslinked PEI micro/nanoparticle (with diameters ranging from 400 nm to 3 μm) in membrane casting solution using epichlorohydrin as crosslinker.

3.2. Surface modification of porous membrane by chemical reaction

Pore-filling method by chemical reaction is another strategy to produce a tight structure on the membrane surface (Figure 3). This method has been used in commercial membrane fabrication to modify the characteristic of membrane or covers the defective pore in membrane surfaces structure 87, 88]. A porous membrane support is immersed in a polymer solution containing other type of polymer or particles. Further thermal treatment is performed to form cross-link structure and then dried at room temperature until all solvents are evaporated. The filled membrane is immersed in a solution to extract the unwanted component from the surface structure. Lan et al. [71] filled polyvinylidene fluoride (PVDF) microfiltration membranes with phenolic resins (PRs)/block copolymers (BCPs). They dipped the PVDF membrane in a homogenous solution containing resol and F127 polymer. The coating membrane was heated to cross-link the resol and generate thermosetting PRs, while F127 was dispersed in PRs structure. The pore-filled membrane was immersed in an acid solution (H_2SO_4) to remove the F127, thus a mesopores surface layer was formed with a pore size of 2350 Da.

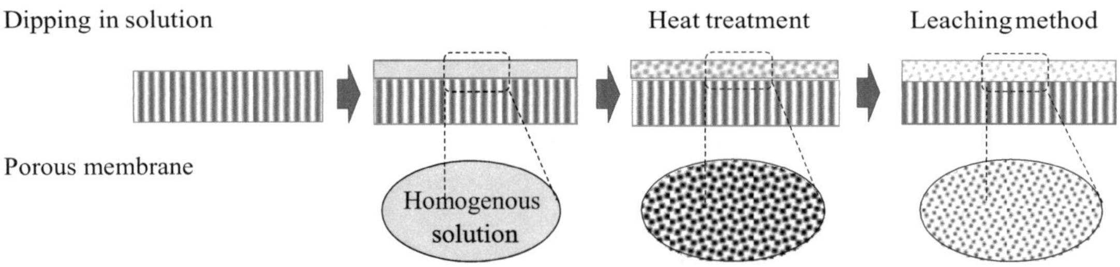

Figure 3. Schematic illustration of surface modification of porous membrane by pore-filling method.

UV-initiated graft copolymerization of poly(ethylene glycol) methacrylate (PEGMA) onto a commercial tight UF PES membranes has been performed for humic acid removal 89]. Smaller membrane pore size was produced when the energy of UV irradiation was between 5 to 9 J/cm^2. Further increase of UV irradiation energy resulted in a degradation of membrane pore leading to a higher membrane pore size. Another method was proposed by Chen et al. [90] to modify the surface of α-alimina UF by TiO_2 nanoparticle dopping sol-gel method. The resulted UF membrane had a pore size with pore size approaching the tight UF range. Several techniques for tight UF preparations is summarized in Table 2.

4. Challenges of Tight UF Membrane for Drinking Water Treatment

4.1. Trade-off between selectivity and permeability.

The most problem in tight UF membrane fabrication is trade-off between selectivity and permeability. These tight UF membranes had too low permeability and higher operating pressures than conventional UF membranes. Ariono et al. [65] found that higher and tighter of membrane pore size was obtained when the polymer concentration was increased, while the permeability was sharply reduced (Figure 4). The increase of polymer concentration increased the viscosity of the membrane solution, which inhibited the growth of membrane pore and reduced the pure water flux. In contrast, the tight membrane skin structure increaseed the membrane rejection. Further developments are continued to improve the membrane permeability while retaining tight pores [91].

Table 2. Several techniques in tight UF membrane preparation.

Modification Techniques	Methods	Membrane pore size	References
Blending	Blending polyethersulfone with PVP (1300 Da)	2000 Da	[74]
	Blending polysulfone with PEG400 and acetone	0.01 μm or 10 nm	[75]
In-situ generation of fillers	In-situ generation of crosslinked PEI in PVDF membrane structure	-	[72]
Dip-coating followed by leaching method	Supported PVDF filled with resin/polymer solution and followed by acid treatment using H_2SO_4 solution	2350 ± 200	[71]
Nanoparticle doping sol-gel method	α-alumina microfiltration filled with TiO_2 nanoparticle (average particle size: 25 nm)	14.700	[90]
UV-initiated graft copolymerization	Graft copolymerization of poly(ethylene glycol) methacrylate (PEGMA) onto a commercial tight UF PES membranes.	4 – 10	[89]

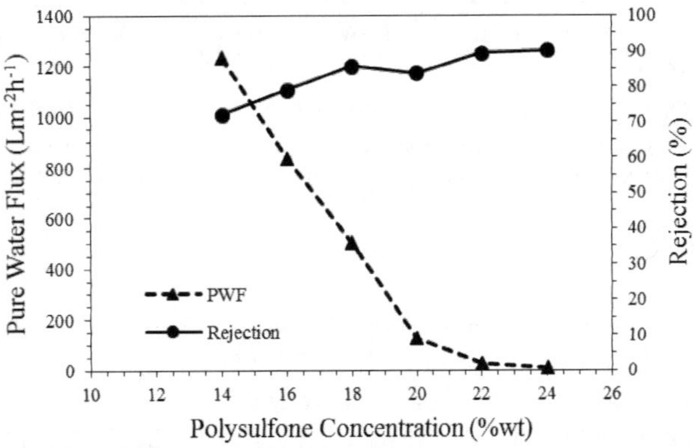

Figure 4. Trade-off between selectivity and permeability during tight UF membrane preparation [64].

4.2. Fouling of organic matter during drinking water treatment

Fouling phenomena have hindered the application of tight UF polymeric membranes for the proposed one-step drinking water disinfection. Although negatively charged tight UF membranes have shown some fouling resistance to colloids and humic materials, these membranes are particularly susceptible to chemical degradation due to chlorine and another oxidant attack. Therefore, developments of commercially viable fouling resistant polymeric membranes have been conducted to achieve one-step disinfection without the need for chemical pretreatment. Several techniques to produce anti-fouling UF membranes have been proposed, from passive anti-fouling to active anti-fouling [91-92]. The passive anti-fouling strategies are developed by manipulating the physicochemical or topological structure of the membrane surface to reduce the interaction between foulant and the membrane surface. Introducing hydrophilic polymers (for examples: PEG-based, PVP-based, and zwitterionic) through hydrogen bonding or ionic solvation is considered as an effective method to prevent the adsorption of foulants on the membrane surface[93,94].

In addition to the hydrophilic polymers, some anti-microbial inorganic nanoparticles (NPs) are also used to avoid the adsorption of foulants to the membrane surface, such as metals (silver/Ag) and metal oxides (TiO_2, SiO_2, Fe_3O_4, and ZnO). Among these NPs, Ag NPs are commonly used as anti-fouling agent in the membrane structure due to its ability to disrupt the cell of bacteria and kill the bacteria when attached on the membrane surface. When using NPs in UF membrane modification for drinking water treatment, the most problems that need to be addressed is the release of NPs from the membrane structure to the environment. Up to this time, a few strategies have been proposed to improve the entrapment of NPs in membrane structure, such as PVP or PEG addition into the membrane solution during membrane preparation and modification of the membrane surface before depositing silver on the surface.

5. Conclusions

Some techniques for preparation of tight UF membrane have been developed to answer the challenges of conventional UF membranes. The tight UF membrane which has a tight structure in the skin layer showed a high rejection towards the TrOCs and low-irreversible fouling compared to the conventional UF membranes. In some cases, the tight UF membrane can be used as an alternative to NF membranes.

Two methods have been proposed to fabricate the low-MWCO UF membranes, i.e. blending nanoparticles or volatile solvent into membrane solution and modifying membrane surface by chemical or plasma grafting method. Low water flux and high operating pressure are challenges of the tight UF membrane application in water treatment which need to be solved. Hence, the development of tight UF membrane has focused on improving flux while maintaining high selectivity.

Introducing hydrophilic polymers or anti-fouling agents onto the membrane surface is considered as an effective method to improve the membrane permeability as well as to avoid the adsorption of foulants to the membrane surface. Recently, nanoparticles (NPs) with anti-microbial properties are increasingly used in membrane modification to reduce fouling formation on the membrane surface. When using the nanoparticles in UF membrane modification for drinking water treatment, the main problem is the release of NPs from the membrane structure to the environment. Therefore, a few strategies should be considered to improve the entrapment of NPs in membrane structure.

6. References

[1] Sirkar KK, Fane AG, Wang R, Wickramasinghe SR. 2015 Process intensification with selected membrane processes *Chemical Engineering and Processing: Process Intensification* **87**:16- 25

[2] Himma NF, Prasetya N, Anisah S, Wenten IG 2018 Superhydrophobic membrane: progress in preparation and its separation properties *Reviews in Chemical Engineering* 20170030

[3] Wenten IG, Khoiruddin K, Hakim AN, Himma NF 2017 The Bubble Gas Transport Method *Part 3: Physical and Chemical Characterization Methods for Membrane Characterization, (Membrane Characterization)* ed Nidal Hilal et.al., (Amsterdam: Elsevier) Chapter 11 pp. 199-218

[4] Du X, Wang Y, Leslie G, Li G, Liang H 2016 Shear stress in a pressure-driven membrane system and its impact on membrane fouling from a hydrodynamic condition perspective: a review *J Chem Technol Biotechnol* **92**(3) 463-478

[5] Wenten IG, Ariono D, Purwasasmita M, Khoirudin 2017 Integrated processes for desalination and salt production: A mini-review *AIP Conference Proceedings* **1818**(1) p. 020065

[6] Wardani AK, Hakim AN, Khoiruddin, Destifen W, Goenawan A, Wenten IG 2017 Study on the influence of applied voltage and feed concentration on the performance of electrodeionization in nickel recovery from electroplating wastewater. *AIP Conference Proceedings* **1805**(1) p. 030004

[7] Wardani AK, Hakim AN, Khoiruddin, Wenten IG 2017 Combined ultrafiltration-electrodeionization technique for production of high purity water *Water Science and Technology* **75**(12) 2891-9

[8] Khoiruddin, Aryanti PTP, Hakim AN, Wenten IG 2017 The role of ion-exchange membrane in energy conversion. *AIP Conference Proceedings* **1840**(1) 090006

[9] Basile A, Cassano A, Rastogi NK 2015 Advances in Membrane Technologies for Water Treatment: Materials, Processes and Applications in *Woodhead Publishing Series in Energy* Elsevier

[10] Bottino A, Capannelli G, Comite A, Ferrari F, Firpo R, Venzano S 2009 Membrane technologies for water treatment and agroindustrial sectors *Comptes Rendus Chimie* **12**(8):882-8

[11] Peters T. 2010 Membrane Technology for Water Treatment *Chemical Engineering & Technology* **33**(8) 1233-40

[12] Quist-Jensen C, Macedonio F, Drioli E 2015 Membrane technology for water production in agriculture: Desalination and wastewater reuse *Desalination* **364** 17-32

[13] Subramani A, Jacangelo JG 2015 Emerging desalination technologies for water treatment: a critical review. Water Research **75** 164-87

[14] Shi X, Tal G, Hankins NP, Gitis V 2014 Fouling and cleaning of ultrafiltration membranes: a review *Journal of Water Process Engineering* **1** 121-38

[15] Aryanti PTP, Sianipar M, Zunita M, Wenten IG 2017 Modified membrane with antibacterial properties *Membrane Water Treatment* **8**(5) 463-81

[16] Sianipar M, Kim SH, Khoiruddin, Iskandar F, Wenten IG 2017 Functionalized carbon nanotube (CNT) membrane: progress and challenges *RSC Advances* **7**(81) 51175-98

[17] Yüksel S, Kabay N, Yüksel M. 2013 Removal of bisphenol A (BPA) from water by various nanofiltration (NF) and reverse osmosis (RO) membranes *Journal of hazardous materials* **263** 307-10

[18] Al-Rifai JH, Khabbaz H, Schäfer AI. 2011 Removal of pharmaceuticals and endocrine disrupting compounds in a water recycling process using reverse osmosis systems *Separation and Purification Technology* **77**(1) 60-7

[19] Wenten IG, Khoiruddin 2016 Reverse osmosis applications: Prospect and challenges *Desalination* **391**112-25

[20] Kozisek F 2005 Health risks from drinking demineralised water *Nutrients in Drinking Water* (Geneva : Sanitation and Health Protection and the Human Environment, World Health Organization) p.148-63

[21] Verma KC, Kushwaha AS 2014 Demineralization of drinking water: Is it prudent? *Medical Journal Armed Forces Indi*a **70**(4) 377-9

[22] Donohue J.M., Abernathy C.O., Lassovszky P. and Hallberg G 2005 The contribution of drinking water to total daily dietary intakes of selected trace mineral nutrients in the United States. In *Nutrients in drinking water* (World Health Organization, Ed. Geneva) pp 75-91

[23] Avni N, Eben-Chaime M, Oron G 2013 Optimizing desalinated sea water blending with other sources to meet magnesium requirements for potable and irrigation waters *Water Research* **47**(7) 2164-76

[24] World Health Organization. Nutrients in drinking water. 2005. http://www.who.int/water_sanitation_health/dwq/nutrientsindw.pdf

[25] Loo S-L, Fane AG, Krantz WB, Lim T-T 2012 Emergency water supply: A review of potential technologies and selection criteria *Water research* **46**(10) 3125-51

[26] Zeman LJ, Zydney AL 1996 Microfiltration and ultrafiltration: principles and applications (New York : Dekker)

[27] Li D, Hu J, Low Z-X, Zhong Z, Wang Y 2016 Hydrophilic ePTFE membranes with highly enhanced water permeability and improved efficiency for multipollutant control *Industrial & Engineering Chemistry Research* **55**(10)2806-12

[28] Liang S, Kang Y, Tiraferri A, Giannelis EP, Huang X, Elimelech M 2013 Highly hydrophilic polyvinylidene fluoride (PVDF) ultrafiltration membranes via postfabrication grafting of surface-tailored silica nanoparticles *ACS applied materials & interfaces* **5**(14):6694-703

[29] Aryanti PTP, Yustiana R, Purnama R, Wenten IG 2015 Performance and characterization of PEG400 modified PVC ultrafiltration membrane *Membrane Water Treatment* **6**(5) 379-92

[30] Fujioka T, Khan SJ, McDonald JA, Nghiem LD 2015 Validating the rejection of trace organic chemicals by reverse osmosis membranes using a pilot-scale system *Desalination* **358** 18-26

[31] Geng Z, Yang X, Boo C, Zhu S, Lu Y, Fan W, et al. 2017 Self-cleaning anti-fouling hybrid ultrafiltration membranes via side chain grafting of poly (aryl ether sulfone) and titanium dioxide *Journal of Membrane Science* **529** 1-10

[32] Van der Bruggen B 2009 Chemical modification of polyethersulfone nanofiltration membranes: a review *Journal of Applied Polymer Science* **114**(1) 630-42

[33] Miller D, Dreyer D, Bielawski C, Paul D, Freeman B 2016 Surface Modification of Water Purification Membranes: a Review *Angewandte Chemie International Edition* **56**(17) 4662-4711

[34] Lalia BS, Kochkodan V, Hashaikeh R, Hilal N 2013 A review on membrane fabrication: Structure, properties and performance relationship *Desalination* **326** 77-95

[35] Lau WJ, Ismail AF, Matsuura T, Nazri NA, Yuliwati E 2015 Advanced Materials in Ultrafiltration and Nanofiltration Membranes *Handbook of Membrane Separations: Chemical, Pharmaceutical, Food and Biotechnological Applications* (New York: CRC Press) p. 7-34

[36] Jhaveri JH, Murthy Z 2016 A comprehensive review on anti-fouling nanocomposite membranes for pressure driven membrane separation processes *Desalination* **379** 137-54

[37] Kumar RS, Arthanareeswaran G, Paul D, Kweon JH 2015 Modification methods of polyethersulfone membranes for minimizing fouling–Review *Membrane Water Treatment* **6**(4) 323-37

[38] Kaner P, Rubakh E, Kim DH, Asatekin A 2017 Zwitterion-containing polymer additives for fouling resistant ultrafiltration membranes *Journal of Membrane Science* **533** 141-59

[39] Bu-Rashid KA, Czolkoss W 2007 Pilot Tests of Multibore UF Membrane at Addur SWRO Desalination Plant, Bahrain *Desalination* **203**(1) 229-42

[40] Heijnen M, Winkler R, Berg P 2012 Optimisation of the geometry of a polymeric Multibore® ultrafiltration membrane and its operational advantages over standard single bore fibres *Desalination and Water Treatment* **42**(1-3) 24-9

[41] Kabay N, Köseoğlu P, Yavuz E, Yüksel Ü, Yüksel M 2013An innovative integrated system for boron removal from geothermal water using RO process and ion exchange-ultrafiltration hybrid method *Desalination* **316** 1-7

[42] Zhenjiang L 2010 The innovation of seven-channel ultrafiltration hollow fiber membranes made from PVC alloy [J]. *Membrane Science and Technology* **4** 027

[43] Bennett A. 2012 Membrane technology: Developments in ultrafiltration technologies *Filtration and Separation* **49**(6) 28-33

[44] Wenten IG. Instalasi unit ultrafiltrasi untuk penyediaan air minum di Desa Cisarua, Kabupaten Purwakarta, Jawa Barat. Report of community service activity 2016: LPPM - Institut Teknologi Bandung. 2016.

[45] He Y 2009 Transportable membrane system produces drinking water. *Membrane Technology* **2009**(8) 8-9

[46] Barbot E, Carretier E, Wyart Y, Marrot B, Moulin P. 2009 Transportable membrane process to produce drinking water *Desalination* **248**(1-3) 58-63

[47] GDP Filter Indonesia 2017 IGW Emergency UF (Portable Hand Pump): G.D.P Filter Indonesia; [17th April]. Available from: http://gdpfilter.co.id/.

[48] Li X, Rao Z, Yang Z, Guo X, Huang Y, Zhang J, et al. 2015 A Survey of 42 Semi-Volatile Organic Contaminants in Groundwater along the Grand Canal from Hangzhou to Beijing, East China *International journal of environmental research and public health* **12**(12) 16070- 81

[49] Vandanjon L, Johannsson R, Derouiniot M, Bourseau P, Jaouen P. 2007 Concentration and purification of blue whiting peptide hydrolysates by membrane processes *Journal of Food Engineering* **83**(4) 581-9
[50] Soffer Y, Aim RB, Adin A. 2005 Membrane fouling and selectivity mechanisms in effluent ultrafiltration coupled with flocculation *Water Science & Technology* **51**(6-7) 123-34
[51] Ellouze F, Ben Amar N, Mokhtar MN, Zimmermann W, Deratani A. 2011 Fractionation of homologous CD6 to CD60 cyclodextrin mixture by ultrafiltration and nanofiltration *Journal of Membrane Science* **374**(1–2) 129-37
[52] Pearce G 2007 Introduction to membranes: filtration for water and wastewater treatment *Filtration & separation* **44**(2) 24-7
[53] Aryanti PTP, Subagjo S, Ariono D, Wenten IG 2015 Fouling and rejection characteristic of humic substances in polysulfone ultrafiltration membrane *Journal of Membrane Science and Research* **1** 41-5
[54] Lin J 2015 Membrane Technologies for Fractionation of Dye/Salt Mixture and Resource Reuse in Textile Industry (Leuven: Katholieke Universiteit Leuven)
[55] Kramer FC, Shang R, Heijman SG, Scherrenberg SM, van Lier JB, Rietveld LC 2015 Direct water reclamation from sewage using ceramic tight ultra-and nanofiltration *Separation and Purification Technology* **147** 329-36
[56] Lin J, Ye W, Baltaru M-C, Tang YP, Bernstein NJ, Gao P, et al 2016 Tight ultrafiltration membranes for enhanced separation of dyes and Na_2SO_4 during textile wastewater treatment *Journal of Membrane Science* **514** 217-28
[57] Ma X, Chen P, Zhou M, Zhong Z, Zhang F, Xing W 2017 Tight ultrafiltration ceramic membrane for separation of dyes and mixed salts (both NaCl/Na2SO4) in textile wastewater treatment *Industrial & Engineering Chemistry Research* **56**(24) 7070–7079
[58] Efligenir A, Déon S, Fievet P, Druart C, Morin-Crini N, Crini G 2014 Decontamination of polluted discharge waters from surface treatment industries by pressure-driven membranes: removal performances and environmental impact *Chemical Engineering Journal* **258** 309-19
[59] Shang R, Verliefde AR, Hu J, Zeng Z, Lu J, Kemperman AJ, et al. 2014 Tight ceramic UF membrane as RO pre-treatment: The role of electrostatic interactions on phosphate rejection. *Water Research* **48** 498-507
[60] Shang R, Verliefde AR, Hu J, Heijman SG, Rietveld LC 2014 The impact of EfOM, NOM and cations on phosphate rejection by tight ceramic ultrafiltration. *Separation and Purification Technology* **132** 289-94
[61] Park N, Lee Y, Lee S, Cho J. 2007 Removal of taste and odor model compound (2, 4, 6-trichloroanisole) by tight ultrafiltration membranes *Desalination* **212**(1) 28-36
[62] Lin Y, Cai Y, Qiu M, Drioli E, Fan Y 2015 Environment-benign preparation of Ag toughening TiO2/Ti tight ultrafiltration membrane via aqueous sol–gel route. *Journal of Materials Science* **50**(15) 5307-17
[63] Wang J, Lang W-Z, Xu H-P, Zhang X, Guo Y-J 2015 Improved poly(vinyl butyral) hollow fiber membranes by embedding multi-walled carbon nanotube for the ultrafiltrations of bovine serum albumin and humic acid *Chemical Engineering Journal* **260** 90-8
[64] Vatanpour V, Madaeni SS, Khataee AR, Salehi E, Zinadini S, Monfared HA 2012 TiO 2 embedded mixed matrix PES nanocomposite membranes: influence of different sizes and types of nanoparticles on antifouling and performance *Desalination* **292** 19-29
[65] Ariono D, Aryanti PTP, Subagjo S, Wenten IG 2017 The effect of polymer concentration on flux stability of polysulfone membrane *AIP Conference Proceedings* **1788**(1) 030048
[66] Setiawan L, Wang R, Li K, Fane AG 2011 Fabrication of novel poly (amide–imide) forward osmosis hollow fiber membranes with a positively charged nanofiltration-like selective layer *Journal of Membrane Science* **369**(1) 196-205
[67] Himma NF, Wardani AK, Wenten IG 2016 Preparation of Superhydrophobic Polypropylene Membrane Using Dip-Coating Method: The Effects of Solution and Process Parameters *Polymer-Plastics Technology and Engineering* **56**(2) 184-194

[68] Giménez-Pérez A, Bikkarolla SK, Benson J, Bengoa C, Stüber F, Fortuny A, et al. 2016 Synthesis of N-doped and non-doped partially oxidised graphene membranes supported over ceramic materials *Journal of Materials Science* **51**(18) 8346-60

[69] Li J-F, Xu Z-L, Yang H, Yu L-Y, Liu M 2009 Effect of TiO 2 nanoparticles on the surface morphology and performance of microporous PES membrane *Applied Surface Science* **255**(9) 4725-32

[70] Wang H, Zhao X, He C 2016 Enhanced antifouling performance of hybrid PVDF ultrafiltration membrane with the dual-mode SiO2-g-PDMS nanoparticles *Separation and Purification Technology* **166** 1-8

[71] Rajaeian B, Heitz A, Tade MO, Liu S 2015 Improved separation and antifouling performance of PVA thin film nanocomposite membranes incorporated with carboxylated TiO 2 nanoparticles *Journal of Membrane Science* **485** 48-59

[72] Lan Q, Yan N, Wang Y 2017 Tight ultrafiltration membranes of mesoporous phenolic resin filled in macroporous substrates *Journal of Membrane Science* **533** 96-102

[73] Kotte MR, Cho M, Diallo MS 2014 A facile route to the preparation of mixed matrix polyvinylidene fluoride membranes with in-situ generated polyethyleneimine particles. *Journal of Membrane Science* **450** 93-102

[74] Karim Z, Claudpierre S, Grahn M, Oksman K, Mathew AP 2016 Nanocellulose based functional membranes for water cleaning: Tailoring of mechanical properties, porosity and metal ion capture *Journal of Membrane Science* **514** 418-28

[75] Sadrzadeh M, Bhattacharjee S 2013Rational design of phase inversion membranes by tailoring thermodynamics and kinetics of casting solution using polymer additives *Journal of Membrane Science* **441** 31-44

[76] Aryanti PTP, Joscarita SR, Wardani AK, Subagjo S, Ariono D, and Wenten IG 2016 The Influence of PEG400 and Acetone on Polysulfone Membrane Morphology and Fouling Behaviour *Journal of Engineering and Technological Sciences* **48**(2) 135-49

[77] Barth C, Goncalves M, Pires A, Roeder J, Wolf B 2000 Asymmetric polysulfone and polyethersulfone membranes: effects of thermodynamic conditions during formation on their performance *Journal of Membrane Science* **169**(2) 287-99

[78] Aryanti PTP, Khoiruddin, and Wenten IG 2013 Influence of additives on polysulfone-based ultrafiltration membrane performance during peat water filtration, *Journal of Water Sustainability* **3**(2) 85-96

[79] Hamid N, Ismail AF, Matsuura T, Zularisam A, Lau WJ, Yuliwati E, et al. 2011 Morphological and separation performance study of polysulfone/titanium dioxide (PSF/TiO 2) ultrafiltration membranes for humic acid removal *Desalination* **273**(1) 85-92

[80] Leo C, Kamil NA, Junaidi M, Kamal S, Ahmad A 2013 The potential of SAPO-44 zeolite filler in fouling mitigation of polysulfone ultrafiltration membrane *Separation and Purification Technology* **103** 84-91

[81] Maximous N, Nakhla G, Wan W, Wong K. 2010 Performance of a novel ZrO_2/PES membrane for wastewater filtration *Journal of Membrane Science* **352**(1) 222-30

[82] Shen J-n, Ruan H-m, Wu L-g, Gao C-j 2011 Preparation and characterization of PES–SiO_2 organic–inorganic composite ultrafiltration membrane for raw water pretreatment *Chemical Engineering Journal* **168**(3) 1272-8

[83] Han R, Zhang S, Liu C, Wang Y, Jian X 2009 Effect of NaA zeolite particle addition on poly (phthalazinone ether sulfone ketone) composite ultrafiltration (UF) membrane performance *Journal of Membrane Science* **345**(1) 5-12

[84] Leo C-P, Lee WC, Ahmad AL, Mohammad AW 2012 Polysulfone membranes blended with ZnO nanoparticles for reducing fouling by oleic acid *Separation and purification technology* **89** 51-6

[85] Balta S, Sotto A, Luis P, Benea L, Van der Bruggen B, Kim J 2012 A new outlook on membrane enhancement with nanoparticles: the alternative of ZnO *Journal of membrane science* **389** 155-61

[86] Taurozzi JS, Arul H, Bosak VZ, Burban AF, Voice TC, Bruening ML, et al. 2008 Effect of filler incorporation route on the properties of polysulfone–silver nanocomposite membranes of different porosities *Journal of Membrance Science* **325**(1) 58-68

[87] Mollahosseini A, Rahimpour A, Jahamshahi M, Peyravi M, Khavarpour M 2012 The effect of silver nanoparticle size on performance and antibacteriality of polysulfone ultrafiltration membrane *Desalination* **306** 41-50

[88] Kim YS, Yang SM 2002 Preparation of continuous mesoporous silica thin film on a porous tube *Advanced Materials* **14**(15) 1078-81

[89] Jin Z, Yiqun F, Nanping X 2010 Preparation and characterization of alumina membranes on capillary supports: effect of film-coating on crack-free membrane preparation *Chinese Journal of Chemical Engineering* **18**(3) 377-83

[90] Peeva PD, Palupi AE, Ulbricht M. 2011 Ultrafiltration of humic acid solutions through unmodified and surface functionalized low-fouling polyethersulfone membranes – Effects of feed properties, molecular weight cut-off and membrane chemistry on fouling behavior and cleanability *Separation and Purification Technology* **81**(2) 124-33

[91] Chen X, Lin Y, Lu Y, Qiu M, Jing W, Fan Y 2015 A facile nanoparticle doping sol–gel method for the fabrication of defect-free nanoporous ceramic membranes *Colloids and Interface Science Communications* **5**(Supplement C) 12-5

[92] Ariono D, Aryanti PTP, Hakim AN, Subagjo S, Wenten IG 2017 Determination of thermodynamic properties of polysulfone/PEG membrane solutions based on Flory-Huggins model *AIP Conference Proceedings* **1840** (1) 090008

[93] Zhang R, Liu Y, He M, Su Y, Zhao X, Elimelech M, Jiang Z 2016 Antifouling membranes for sustainable water purification: strategies and mechanisms *Chemical Society Reviews* **45**(21) 5888-924

[94] Aryanti PTP, Subagjo S, Ariono D, Wenten IG. 2015 Fouling and rejection characteristic of humic substances in polysulfone ultrafiltration membrane. *Journal of Membrane Science and Research* **1**(1) 41-45

[95] Wenten IG, Aryanti PTP, Khoiruddin K, Hakim AN, Himma NF. 2016 Advances in Polysulfone-Based Membranes for Hemodialysis *Journal of Membrane Science and Research* **2**(2)78-89

IOP Conf. Series: Materials Science and Engineering **395** (2018) 012013 doi:10.1088/1757-899X/395/1/012013

Transversal Submerged Membrane Contactor for Simultaneous Absorption and Desorption of CO_2 During Natural Gas Sweetening Process

U W R Siagian[1], Y Rahmawati[2], K Khoiruddin[3], A N Hakim[3], I G Wenten[3,*]

[1] Department of Petroleum Engineering, Institut Teknologi Bandung, Jl. Ganesha 10, Bandung 40132, Indonesia

[2] Department of Chemical Engineering, Institut Teknologi Sepuluh Nopember, Surabaya, Indonesia

[3] Department of Chemical Engineering, Institut Teknologi Bandung, Jl. Ganesha 10, Bandung 40132, Indonesia

*Corresponding author's e-mail: igw@che.itb.ac.id

Abstract. In this work, a unique transversal submerged membrane contactor is used for CO_2 removal. The developed configuration is aimed to provide the possibility to perform simultaneous absorption and desorption of carbon dioxide in a single module. In addition, this configuration is expected to reduce the solvent circulation which is the major obstacle encountered in separated absorption-desorption unit. The module contains a number of polypropylene hollow fiber membranes arranged perpendicularly to the longitudinal axis in which the fibers are submerged into a solvent. Pure water which is placed in the shell of the module is used as the solvent. The effects of operating parameters such as gas flow rate (40-80 ml/min) and operating time on membrane performance are investigated. In addition, the simultaneous absorption-desorption process is compared to the separated ones. Results revealed that CO_2 removal is significantly affected by the gas flow rate. The optimum gas flow rate is 60 ml/min (vacuum pressure of 60.5 kPa) with >80% CO_2 absorption and desorption efficiencies. During a long-term operation, the separation efficiencies are decreased after 200 minutes of operation which may be associated with membrane wetting phenomenon. The simultaneous absorption-desorption process shows a higher separation efficiency rather than the separated process.

1. Introduction

In natural gas sweetening process, CO_2 should be reduced to a certain level to prevent several problems including a freezing in low-temperature chillers, corrosion, reduction of the caloric value, and reduction of pipeline capacity [1]. There are several methods that have been used to remove CO_2 such as solvent-based absorption, solid adsorption, and membrane-based separation. Absorption using amine-based solvent is currently the well-established method. It also has a high selectivity and a high absorption capacity [2,3]. However, conventional absorption column has several disadvantages such as high energy for solvent regeneration, difficulty designing the gas-liquid mass transfer area, and a limited range of gas and liquid flow rates due to operational problems e.g. flooding, loading, channeling, and foaming [4,5]. Membrane technology has been considered as the attractive alternative and has started to replace the conventional technologies in various processes including separation, chemical reaction, and contacting processes [6–14]. Membrane offers advantages such as relatively lower energy consumption, relatively low operating cost, smaller footprint, easy to scale-up, and easy to operate which make the rapid development and applications of membrane in almost industrial sector [15–20]. In addition, membrane technology also provides a more intensive process and can be combined with several processes into a hybrid process thus the processing steps can be reduced into a single unit [21–25]. Membrane contactor is a hybrid process that combines the advantages of selective absorbent with membrane contactor. The use of membrane as contacting device is aimed to solve the operational problems encountered in conventional absorption column. However, in a separated adsorption and desorption process, circulation of a large volume of solvent is needed for a continuous process which will increase the operational expenditure due to the high cost of solvent. Therefore, development of new configuration is needed to solve this problem.

In this study, a new configuration of transversal submerged membrane contactor is proposed to overcome operational problems of the separated adsorption and desorption process for CO_2 removal. A transversal module is used to improve the separation performance of membrane contactor. Submerged membrane configuration is expected to avoid solvent recirculation and reduce the operating cost. Moreover, a simultaneous desorption or solvent regeneration is aimed to improve mass transfer rate during the absorption process.

2. Materials and method

Hollow fiber polypropylene (PP) membrane (supplied by GDP Filter, Indonesia) was used to fabricate membrane module. The PP fiber has 0.35 mm inner diameter, 0.5 mm outer diameter, and 0.2 μm pore size. The total surface area of the membrane was 0.27 m^2. Pure CO_2 gas (obtained from a local supplier) was used as the feed gas. Pure water was used as solvent (absorbent).

Experimental set up used in this study is shown in Figure 1. The set-up consists of CO_2 vessel, membrane module, trap, and vacuum pump (Figure 1 (a)). Pure water is placed in the shell of the module wherein the membranes are submerged into the solvent. The module contains of hollow fiber PP which is arranged perpendicularly to longitudinal axis of the module (Figure 1 (b)). Moreover, the fibers comprise of two layers for feed and permeate gas (Figure 1 (c)). Pure CO_2 gas is transferred into PP fiber lumen of the firs layer at 40-80 ml/min (Figure 1 (d)). Then, the gas diffuses into the shell side of the module via membrane pores. In the shell side, the gas is absorbed by the solvent. Simultaneously, the absorbed CO_2 is desorbed through the adjacent fiber or the second layer by applying vacuum pressure (60.5 kPa) (Figure 1 (d)). In this way, the simultaneous adsorption-desorption process occurs. The amount of CO_2 absorbed in the solvent is analyzed by titration method using sodium hydroxide solution (0.1 M). Performance of the membrane contactor is analyzed based on CO_2 flux and removal efficiency.

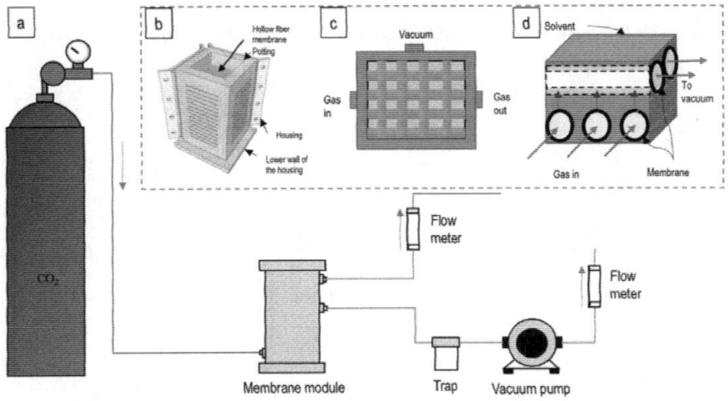

Figure 1. Experimental set-up. (a) Schematic of CO_2 removal experiments, (b) transversal membrane contactor, (c) top view of transversal membrane contactor, and (d) absorption-desorption process in submerged transversal membrane contactor.

3. Results and discussion

Performances of membrane contactor in absorption and desorption processes are shown in Figure 2 (a) and (b). It can be observed that CO_2 removal is increased when the gas flow rate is increased from 40 to 60 ml/min. The increasing CO_2 removal may be associated with a decreasing boundary layer thickness which leads to a better gas diffusion from membrane side to the solvent in the shell side. However, a further increase of the gas flow rate results in a reduction of CO_2 removal. This is due to a shorter residence time of gas in the membrane which reduces the transport of CO_2 from gas phase to the liquid phase. Meanwhile, CO_2 flux increases with the increase of gas flow rate. This may be due to the fact that the amount of CO_2 increases as the gas flow rate is increased. In desorption, both CO_2 flux and separation decrease along with increasing gas flow rate. The decreasing desorption efficiency may be associated with the increasing amount of CO_2 absorbed in the solvent. Since the amount of absorbed CO_2 is higher while the vacuum pressure is constant, the amount of desorbed CO_2 is almost the same. Consequently, the desorption rate is reduced. In addition, the declining CO_2 flux at higher gas flow rate may be attributed to the reduction of the gas film layer on the membrane surface. This will reduce the mass transfer rate during the desorption process.

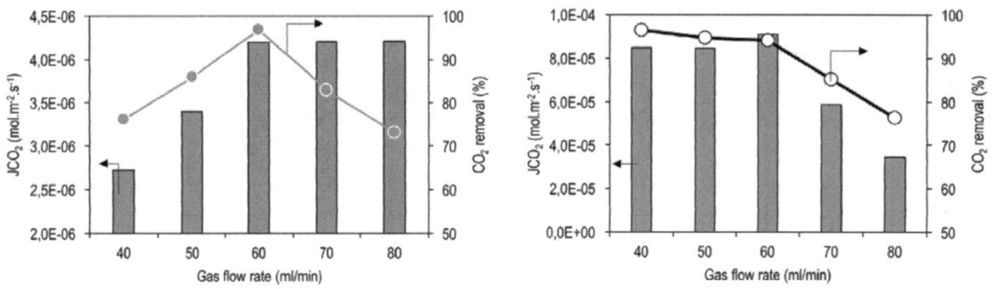

(a) (b)

Figure 2. The effect of gas flow rate on CO_2 flux and removal efficiency during (a) absorption and (b) desorption process.

Figure 3 shows the performance of membrane contactor at the various operating time. The performance of the membrane is relatively stable during the first 200 min. The performance of membrane declines both in absorption and desorption processes after 200 min operating time. The declining absorption rate (CO_2 removal and flux) may be due to a wetting phenomenon. Wetting gives additional mass transfer resistance which reduces or restricts the diffusion of gas through the membrane. Therefore, the amount of CO_2 absorbed and the CO_2 flux decrease during this period. The wetting is proved by the result of CO_2 concentration analysis in the solvent which is still below the maximum solubility. Since the adsorption and desorption occur simultaneously, the declining absorption rate also leads to decreasing desorption rate. Therefore, during this period, the desorption performance also deteriorates along with declining absorption performance.

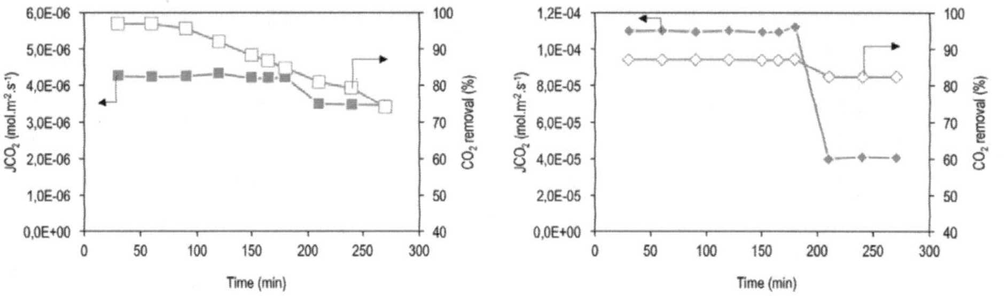

(a) (b)

Figure 3. Flux and removal efficiency of (a) absorption and (b) desorption process at various operating time.

The performance of simultaneous adsorption-desorption process in the transversal submerged membrane contactor is compared to the separated process. Table 1 shows the comparison of CO_2 removal in separated and simultaneous adsorption-desorption process in a submerged transversal membrane contactor obtained in this study. It can be observed that the simultaneous process provides a better CO_2 absorption and desorption efficiency. Since the solvent is continuously regenerated, the solvent is expected to be in unsaturated condition. Accordingly, the mass transfer rate is better. Therefore, the simultaneous process gives a better separation efficiency rather than the separated one. Furthermore, since the absorbent is stagnant in the membrane module, it gives a sufficient residence time for the absorbent in the module. The longer contact time between the membrane and the absorbent provides the opportunity for the absorbent to achieve a higher absorption capacity. Therefore, a maximum CO_2 removal rate can be achieved. In addition, this also provides the sufficient time for absorbent regeneration process. As the result, both CO_2 absorption and desorption in the simultaneous process are improved. In addition, this novel configuration can eliminate solvent recirculation which is one of the disadvantage of conventional absorption process. Also, another advantage of this configuration is no additional stripping unit is required as encountered in general module [26].

Table 1. Comparison of separated and simultaneous absorption-desorption in transversal submerged membrane contactor.

Process	Operational conditions	CO_2 removal rate (%)
Absorption	Q_{gas} = 70 ml/min; Q_l = 109 ml/min;	46.3
Desorption	Q_l = 204 ml/min; P_{vacuum} = 34.5 kPa;	45.8
Absorption-desorption (submerged membrane)	Q_{gas} = 60 ml/min; P_{vacuum} = 60.5 kPa;	Absorption >80% Desorption > 80%

4. Conclusion
In this work, a unique transversal submerged membrane contactor is used for CO_2 removal. The developed configuration is aimed to provide the possibility to perform simultaneous absorption and desorption of carbon dioxide in a single module. In addition, this configuration is expected to reduce the solvent circulation which is the major obstacle encountered in separated absorption-desorption unit. Results revealed that CO_2 removal is significantly affected by the gas flow rate. The optimum gas flow rate is 60 ml/min (vacuum pressure of 60.5 kPa) with >80% CO_2 absorption and desorption efficiencies. During a long-term operation, the separation efficiencies are decreased after 200 minutes operation which may be associated with membrane wetting phenomenon. The simultaneous absorption-desorption process shows a higher separation efficiency rather than the separated process.

5. References
[1] Boucif N, Corriou J P, Roizard D and Favre E 2012 Carbon dioxide absorption by monoethanolamine in hollow fiber membrane contactors: A parametric investigation *AIChE J.* **58** 2843–55
[2] Wang M, Lawal A, Stephenson P, Sidders J and Ramshaw C 2011 Post-combustion CO_2 capture with chemical absorption: a state-of-the-art review *Chem. Eng. Res. Des.* **89** 1609–24
[3] Aaron D and Tsouris C 2005 Separation of CO_2 from flue gas: a review *Sep. Sci. Technol.* **40** 321–48
[4] Mansourizadeh A and Ismail A F 2009 Hollow fiber gas–liquid membrane contactors for acid gas capture: a review *J. Hazard. Mater.* **171** 38–53
[5] Ahmad A L, Sunarti A R, Lee K T and Fernando W J N 2010 CO_2 removal using membrane gas absorption *Int. J. Greenh. Gas Control* **4** 495–8
[6] Drioli E and Curcio E 2007 Membrane engineering for process intensification: a perspective *J. Chem. Technol. Biotechnol.* **82** 223–7
[7] Strathmann H, Grabowski A and Eigenberger G 2013 Ion-exchange membranes in the chemical

process industry *Ind. Eng. Chem. Res.* **52** 10364–79
[8] Rathore A S and Shirke A 2011 Recent developments in membrane-based separations in biotechnology processes: Review *Prep. Biochem. Biotechnol.* **41** 398–421
[9] Himma N F, Wardani A K and Wenten I G 2017 The effects of non-solvent on surface morphology and hydrophobicity of dip-coated polypropylene membrane *Mater. Res. Express* **4** 54001
[10] Himma N F, Wardani A K and Wenten I G 2017 Preparation of superhydrophobic polypropylene membrane using dip-coating method: The effects of solution and process parameters *Polym. - Plast. Technol. Eng.* **56** 184–94
[11] Purwasasmita M, Kurnia D, Mandias F C, Khoiruddin and Wenten I G 2015 Beer dealcoholization using non-porous membrane distillation *Food Bioprod. Process.* **94** 180–6
[12] Purwasasmita M, Nabu E B P, Khoiruddin and Wenten I G 2015 Non dispersive chemical deacidification of crude palm oil in hollow fiber membrane contactor *J. Eng. Technol. Sci.* **47** 426–46
[13] Ariono D, Purwasasmita M and Wenten I G 2016 Brine effluents: Characteristics, environmental impacts, and their handling *J. Eng. Technol. Sci.* **48** 367–87
[14] Wardani A K, Hakim A N, Khoiruddin and Wenten I G 2017 Combined ultrafiltration-electrodeionization technique for production of high purity water *Water Sci. Technol.* **75** 2891–9
[15] Baker R W and Low B T 2014 Gas separation membrane materials: A perspective *Macromolecules* **47** 6999–7013
[16] Ma L, Dong X, Chen M, Zhu L, Wang C, Yang F and Dong Y 2017 Fabrication and water treatment application of carbon nanotubes (CNTs)-based composite membranes: A review *Membranes* **7** 16
[17] Wenten I G and Khoiruddin 2016 Recent developments in heterogeneous ion-exchange membrane: Preparation, modification, characterization and performance evaluation *J. Eng. Sci. Technol.* **11** 916–34
[18] Wenten I G, Dharmawijaya P T, Aryanti P T P, Mukti R R and Khoiruddin 2017 LTA zeolite membranes: Current progress and challenges in pervaporation *RSC Adv.* **7** 29520–39
[19] Wenten I G, Khoiruddin and Himma N F 2017 Membrane-based carbon capture technology: Challenges and opportunities in Indonesia *Adv. Sci. Lett.* **23** 5768–71
[20] Khoiruddin, Ariono D, Subagjo and Wenten I G 2017 Surface modification of ion-exchange membranes: Methods, characteristics, and performance *J. Appl. Polym. Sci.* **134** 45540
[21] Ariono D, Khoiruddin, Subagjo and Wenten I G 2017 Heterogeneous structure and its effect on properties and electrochemical behavior of ion-exchange membrane *Mater. Res. Express* **4** 24006
[22] Aryanti P T P, Sianipar M, Zunita M and Wenten I G 2017 Modified membrane with antibacterial properties *Membr. Water Treat.* **8** 463–81
[23] Wenten I G, Khoiruddin K, Hakim A N and Himma N F 2017 Chapter 11 - The Bubble Gas Transport Method - Membrane Characterization (Elsevier) pp 199–218
[24] Mangindaan D, Khoiruddin K and Wenten I G 2018 Beverage dealcoholization processes: Past, present, and future *Trends Food Sci. Technol.* **71** 36–45
[25] Sianipar M, Kim S H, Khoiruddin, Iskandar F and Wenten I G 2017 Functionalized carbon nanotube (CNT) membrane: Progress and challenges *RSC Adv.* **7** 51175–98
[26] Mansourizadeh A 2012 Experimental study of CO_2 absorption/stripping via PVDF hollow fiber membrane contactor *Chem. Eng. Res. Des.* **90** 555–62

IOP Conf. Series: Materials Science and Engineering **395** (2018) 012014 doi:10.1088/1757-899X/395/1/012014

Synthesis and Applications of Copper Nanopowder – A Review

S Wahyudi[1,2], S Soepriyanto[1], M Z Mubarok[1], Sutarno[3]

[1] Department of Metallurgical Engineering-Faculty of Mining and Petroleum Bandung Institute of Technology, Bandung, Indonesia
[2] Department of Material & Metallurgical Engineering-Faculty of Engineering Bandung Institute of Science & Technology, Bekasi, Indonesia
[3] Department of Metallurgical Engineering- Faculty of Engineering, University of Jenderal Achmad Yani, Bandung, Indonesia

Email: solehwahyudi@students.itb.ac.id

Abstract. Copper nanopowder demonstrates excellent conductivity, catalytic behaviour, excellent compatibility and antibacterial properties. Copper nanopowder has been used as an essential component in the future nano-devices, a wide range of industrial applications and used as a replacement for more expensive metals. Copper nanopowder properties are strongly influenced by shape, size and size distribution which are often varied by varying the synthetic methods. Accordingly, this review presents different of synthetic methods of copper nanopowder and applications of this nanopowder in different fields.

1. Introduction

The copper nanopowder is a copper powder that has particle sizes below 100 nm and considered as an alternative material other than precious metals such as gold and silver that are more economical wherein pure form has good electrical and heat conductivity properties and are antibacterial. In addition, the advantages possessed by nano-sized powder particles are the value of the ratio of surface area to high volume. These considerations cause copper nanopowder widely used for electronics, magnetic, optical, catalysts, and pigments [1].

The potential needs of nanomaterial show an increasing trend [2]. The constraints faced during the execution of the production of copper nanopowder are the sensitivity of the copper nanopowder to the aqueous environment and the air, which will aggregate and oxidize. The steps taken to overcome these constraints are the conditioning of inert environments using argon and nitrogen gas when packaging or providing a protective layer of polymer or surfactant before packaging [1]. Another obstacle is hazard [2].

Nanoparticles are synthesized by three different methods: chemical, physical, and biological [3]. Each synthesis method has advantages and disadvantages in producing copper nanopowder. The synthesis method was chosen to be applied consider several things including morphology, size distribution, cost, the scale of production and hazard [3]. In this review will be presented about the method of synthesis according to its classification by displaying its advantages and disadvantages along with the characteristics of copper nanopowder produced. Similarly, about the application of copper nanopowder. This review complements the previous review of copper nanopowder by displaying information about copper nanopowder that has been produced by industry and the methods used. Specifications of copper nanopowder that has been produced from several industries will also be delivered.

2. Methods of synthesis

The synthesis methods of copper nanopowder can be done by chemical reduction, microemulsion reduction, sonochemical and sonoelectrochemical, microwave, photochemical, electrochemical, sonoelectrochemical, thermal decomposition, laser ablation, mechanical / ball mill, wire discharge, the electric explosion of wire, and the biological method that used bacteria, fungi, and plants. Table 1 listed the copper nanopowder synthesis methods into the chemical, physical, and biological categories.

Table 1. Copper nanopowder synthesis methods.

Chemical methods	Physical methods	Biological methods
Chemical reduction [38, 39]	Laser ablation [24]	Bacteria [34]
Microemulsion reduction [40, 41]	Mechanical mill [25]	Fungi [35]
Sonochemical [42] &	Pulsed Wire Discharge [27]	Plants [37]
Sonoelectrochemical [43]	Electric Explosion of Wire [29]	
Microwave [15]	Gas evaporation [31]	
Photochemical [17]		
Electrochemical [44]		
Thermal decomposition [22]		

The characteristics of copper nanopowder such as size, morphology, and particle size distribution depending on the applied synthesis method. Table 2 listed the characteristics of copper nanopowder in each method.

2.1. Chemical reduction

The chemical reduction method for the synthesis of copper nanopowder is the easiest, simplest and most commonly used [1]. It is generally carried out in aqueous media using copper salts as a source of copper and reducing agents such as sodium borohydride [4], hydrazine [5], ascorbic acid [6], polyol [7], isopropyl alcohol with Cetyl Trimethyl Ammonium Bromide (CTAB) [8] & glucose [9]. Reverberi et al. in his study used a vanadium reducing agent (+2) salt [10]. In addition to being in aqueous media, other media in the form of the organic solvent was once used in the synthesis of copper nanopowder [11]. The morphology and size of the resulting nanopowder particles can be controlled by selecting the reducing agent and regulating the reaction conditions. The disadvantage of this method is its toxic and flammable chemicals. In addition, the surface of the nanoparticles is easily oxidized, so it is necessary to use inert environmental conditions (nitrogen or argon atmospheres) or surface-active agents to protect the surface of nanoparticles such as surfactants, dissolved polymers, weak acids, etc. known as capping agent [1].

2.2. Microemulsion reduction (colloidal)

The synthesis of copper nanopowder applies a chemical reduction method in an organic compound solvent in the form of microemulsion water in oil (W / O) such as Water / Isooctane / Cyclohexane [12] or water oil (O / W) or water in supercritical carbon dioxide (W / Sc. CO_2). Compared with the chemical reduction method above, the application of microemulsion method can produce copper nanopowder with morphology and size distribution more evenly, but its operation is more expensive and needs to separate the solvent from product [1].

2.3. Sonochemical & Sonoelectrochemical

Synthesis of copper nanopowder can be done by applying sonochemical & sonoelectrochemical method. The working principle is to use ultrasound radiation (frequency 20 kHz to 10 MHz) in the copper salt electrolyte solution. The chemical and physical effects of ultrasound arising from the species, growth, and implosive collapse of bubbles [13]. The sonoelectrochemical is the use of ultrasound radiation on the electrochemical method. Some different arrangements of equipment have been used for the introduction of the ultrasound irradiation into the electrochemical systems [14].

2.4. Microwave
A microwave method is a form of electromagnetic energy in the frequency range between 300 MHz to 300 GHz, in the synthesis of copper nanopowder, can accelerate volumetric heating and kinetics, rapid volumetric heating and kinetics, short reaction periods and increasing yields of products compared to conventional heating methods [1]. The size of the resulting nanoparticles can reach 10 nm [15].

2.5. Photochemical
The photochemical method for synthesis of copper nanopowder utilizes light intensity. Photochemical techniques provide several advantages over conventional chemical means, such as (a) the reduction of metal ions can be carried out without using reducing agents and so avoiding undesired by-products of the reductant, (b) through the choice of suitable wavelength and concentration the rate of reaction can be controlled, (c) the light, which works as a reducing agent, is uniformly distributed in the solution, (d) irradiation can be performed at room temperature [16]. Kapoor and Mukherjee [17] using photochemical (UV light) method with poly (N-vinylpyrrolidone) (PVP) as a stabilizer can produce copper nanopowder with an average size of 15 to 20 nm.

2.6. Electrochemical
Copper nanopowder is produced by an electrochemical method by applying an electric current between the anode and cathode electrodes in the electrolyte solution. The reduction process occurs on the surface of the cathode. The deposition of copper powder on the cathode surface can be obtained at higher overpotentials corresponding to the limiting diffusion current density. The reaction of hydrogen evolution also occurs on the surface of the cathode [18]. The electrochemical method is a simple method. The process is fast and more economical to synthesize copper nanopowder. This method offers a relatively clean, non-toxic, process flow process that can be done at room temperature and environmentally friendly [19]. Kadam [19] used copper sulfate as an electrolytic solution and supplied 2 V of voltage and 1.5 A of current for 30 minutes and successfully synthesized copper nanoparticles with the size around of 24 nm and spherical. The problem that still exists in synthesis with electrochemistry is the presence of ionic interactions and inhomogeneous divisions in particle size grains.

2.7. Thermal decomposition
The synthesis of copper nanopowder is carried out in pressurized containers and controlled temperatures such as autoclaves, where the solvent reaches a temperature above its boiling point. Following the type of solvent, called hydrothermal [20, 21] and solvothermal. Karthik et al. [22] used this method for the synthesis of 50 to 100 nm copper nanoparticles. They use copper (II) succinate precursor and oleylamine as a capping agent.

2.8. Laser ablation
The method consists of ablation of a target by intense laser radiation in a liquid, yielding to an ejection of its constituents and the formation of nanoclusters and nanostructures. Variety of liquids can be used in which the particles remain as a suspension [23]. Moniri et al. [24] produce nanoparticles of average 5 nm copper in a 1.2 J/cm^2 laser fluence.

2.9. Mechanical mill
Milling is a solid-state processing technique for nanoparticle synthesis. The micron-sized material is fed into the milling machine. The types of milling machines commonly used for the synthesis of copper nanoparticles are planetary, vibratory, uniball, and the attritor. This method has ease of operation and low production cost but has a long time for nanoparticle production. The important factors affecting the quality of the final product are the type of mill, milling speed, container, time, temperature, atmosphere, size, and size distribution of the grinding medium, process control agent, the

weight ratio of the ball to powder and extent of filling the vial [1]. Yadav et al. [25] produce a 21 nm copper nanopowder with a planetary ball mill.

2.10. Pulsed Wire discharge (PWD)

When a high-pulsed current is driven through a thin metallic wire, the Joule heating causes electrical energy deposition that can turn the whole wire into plasma. By cooling this plasma with ambient gas, we can obtain a large number of very fine solid particles. This method is not usually used in industry due to the high cost and possible inefficiency for some metals. This method is suitable for metals with high electrical conductivity, which can easily be formed into thin wires [26]. Murai et al. [27] using the PWD method with the energy of 68 J successfully synthesized 0.25 mm diameter copper wire into a copper nanopowder with an average size of 25 nm. He also examined the effect of giving coating oleic acid, and the result was 10 nm smaller than uncoiled and unoxidized for several months (over two months) after the preparation.

2.11. The Electric explosion of wire (EEW)

It is a process of explosive destruction of a metal wire under the action of high current density (more than 10^6 A/cm^2) [28]. The following peculiarities characterize EEW: time of explosion is 10^{-5} to 10^{-8} s; the temperature at the moment of explosion can reach the value more than 104 K; pressure up to 109 Pa; velocity of product recession is from 1 to 5 km/s. The material of the wire transmutes into particles of nano-sized range (10 to 100 nm) following specific conditions. Extremely non-equilibrium conditions of EEW cause some unusual properties of nanopowder [28]. Electro-explosive nanopowder has, as a rule, the spherical form of particles, they are steady against oxidation and sintering at room temperature and characterized by high diffusion activity at the heating [28]. EEW in an inert gas or hydrogen is applied to produce metal powders, alloys, and intermetallic compounds. Whereas EEW in an active chemical environment is applied to produce nanopowder of metal chemical compounds: oxides, nitrides, carbides, etc. [28]. Dash et al. [29] using the EEW method successfully synthesized 0.5 mm copper wire into a copper nanopowder with an average size of 36.34 nm and a spherical shape. The experimental conditions are 3 μF of capacitance, 22 kV of charging voltage, 0.1 MPa of Argon gas ambiance pressure.

2.12. Gas evaporation

Synthesis of metal nanopowder by the gas evaporation method is evaporation with the subsequent condensation of metal using high-intensity electron beams in an atmosphere of inert gas. The simple scalability to the industrial conditions, the possibility of evaporating any material and the high-purity of the product are advantages of particles synthesized from the gas phase [30]. Zavjalov et al. [31] obtain copper nanopowder by substance evaporation with the help of a powerful electron beam with the sizes of particles are limited in range from about 15 to 700 nm.

2.13. Biological method

It has been found that living organisms such as bacteria, fungi, and plants have great potential for the synthesis of metal nanoparticles. This method is easier to control the size of the distribution of nanoparticles synthesized than others and no toxic impact on the environment [32, 33]. Metal compounds usually reduce into their respective nanoparticles because of microbial enzymes or the plant phytochemicals with antioxidant or reducing properties [1]. Varshney et al. [34] used *Pseudomonas stutzeri* to synthesize copper nanoparticles from wastewater generated from electroplating. They have a cubic shape, and the size of nanoparticles produced by this method is 50-150 nm [3, 34]. Pavani et al. [35] used the *Aspergillus fungus* species to synthesize copper nanoparticles. Plants can be used for the synthesis of copper nanopowder through inactivated plant tissue, plant extracts, exudates, gums, and other parts of plants [36]. Lee et al. [37] have been biologically researching for the synthesis of copper nanopowder with a size of 40-100 nm using *Mangolia* plant leaf extract as a reducing agent.

IOP Conf. Series: Materials Science and Engineering **395** (2018) 012014 doi:10.1088/1757-899X/395/1/012014

Table 2. Resume research of each copper nanopowder synthesis method.

Method	Material	Conditions	Morphology	Size	Ref.
Chemical reduction	Solvent: DI Water Precursor: $CuCl_2.2H_2O$ Reducing agent: Ascorbic acid	80°C, 14 h	Spherical	< 2 nm	[38]
Chemical reduction	Solvent: DI Water Precursor: $CuSO_4.5H_2O$ Reducing agent: Ascorbic acid Capping agent: Starch	80°C, 2 h	Cubic	25.19 nm	[39]
Microemulsion reduction	Solvent: Water in TX-100/n-hexanol/cyclohexane Precursor: $CuCl_2$ Reducing agent: $NaBH_4$	RT, 10 min	Spherical	5-15 nm	[40]
Microemulsion reduction	Solvent: Water in n-heptane, alcohol Precursor: $CuCl_2$ Reducing agent: $NaBH_4$ Stabilizer: Bis(ethylhexyl)hydrogen phosphate	25°C, 12 h	Spherical	60 nm	[41]
Sonochemical	Solvent: Water Precursor: $CuSO_4$ Reducing agent: Hydrazine monohydrate Stabilizer: Ethylene glycol		Spherical	108 nm	[42]
Sonoelectrochemical	Solvent: Water Precursor: $CuSO_4.5H_2O$ Stabilizer: poly(N-vinylpyrrolidone)	480 mA/cm^2, 30 min	Spherical	25-60 nm	[43]
Microwave	Solvent: Ethylene glycol Precursor: $CuSO_4.5H_2O$ Reducing agent: $NaH_2PO_2.H_2O$ Stabilizer: poly(N-vinylpyrrolidone)	5 min	Spherical	10 nm	[15]
Photochemical	Solvent: Water Precursor: $CuSO_4$ Stabilizer: poly(N-vinylpyrrolidone)	20°C, 20 min	Spherical	15-20 nm	[17]
Electrochemical	Solvent: Water Precursor: $CuSO_4$	2 V, 1.5 A	Spherical	24 nm	[44]
Thermal decomposition	Solvent: Oleylamine Precursor: Copper (II) succinate Stabilizer: Oleylamine	245°C, 45 min		50-100 nm	[22]
Laser abrasion	Copper metal plate	1.2 J/cm^2 RT, 10 min	Spherical	5 and 16 nm	[24]
Mechanical ball	Copper powder Wet milling medium: Toluene Rectified	40 h		21 nm	[25]
Pulsed wire discharge	Copper wire Coating: Oleic acid	473 K, 10 μF, 5.2 kV, 68 J	Spherical	25 nm	[27]
Electric explosion of wire	Copper wire	0.1 MPa, 3 μF, 22 kV	Spherical	36.34 nm	[29]
Gas evaporation	Cylindrical copper ingot Flux: Argon	5 MW/cm^2	Spherical	15-700 nm	[31]
Biological	Solvent: Water Precursor: $CuSO_4.5H_2O$ Biological entity: Bacteria *pseudomonas stutzeri* from electroplating waste			150 nm	[34]
Biological	Solvent: Water Precursor: $CuSO_4.5H_2O$ Biological entity: Fungi *aspergillus fungus*			50-150 nm	[35]
Biological	Solvent: Water Precursor: $CuSO_4.5H_2O$ Biological entity: *Mangolia* plant leaf extract			40-100 nm	[37]

3. Copper nanopowder characteristics

Copper nanopowder synthesis methods that have been applied in the industry are electrochemical, electrical explosion and laser ablation. Table 3 shows the copper nanopowder specifications that produced in the industry as well as the methods used.

Table 3. The copper nanopowder product specification and methods in industry[a].

Method	Purity Cu (% Metal basis)	Color	APS (nm)	SSA (m^2/g)	Morphology	Density (g/cm^3) Bulk	Density (g/cm^3) True	Other	Ref.
EEM	99.9	Saddle brown	40	10-14	Spherical	0.21	10-14		[45]
ECM	99	Black brown	30	35		0.15-0.35	35	PP	[46]
LSM	99.8	Black	25	30-50	Spherical	0.15-0.35	8.94	CC	[47]

[a] US Research Nano Materials, Inc.
EEM=Electrical Explosion Method; ECM= Electrochemical Method; LSM= Laser Synthesized Method;
APS= Average Particle Size; SSA=Specific Surface Area; PP= Partially Passivated Cu_2O; CC= Carbon Coated.

A common characteristic of copper is a ductile metal that has high thermal and electrical conductivity. Pure copper has a soft and orange-reddish color. The characteristics of the above copper nanopowder product specifications can be determined using a test apparatus to determine the purity level of copper nanopowder, particle size, particle size distribution, morphology and specific gravity.

4. Commercial applications using copper nanopowder

Copper is an alternative metal substitute of precious metals that have high electrical conductivity and heat properties making it ideal for reducing production costs and used in industrial applications for high strength metals and alloys, high thermally conductive materials and capacitor materials. As a nanopowder, the main characteristic is to have a high SSA (Specific Surface Area) so that copper nanopowder can be used in a variety of industrial applications, including anti-biotic, antimicrobial, anti-fungal agent, conductive ink, & conductive paste for printed electronics, sintering additives, lubricant additives, and catalysts. [1, 3]

5. Conclusions

Several methods of synthesis of copper nanopowder have been delivered. Chemical methods are the most commonly used method for the preparation of nanoparticles but still involves toxic materials. Physical methods can produce faster results and do not involve toxic materials but are still costly. While biological methods known as nanoparticle synthesis are environmentally friendly but require a very long time synthesis.

Copper nanopowder has very high electrical and heat conduction, excellent physicochemical properties; good biocompatibility and high surface activity, and therefore is very promising for magnetic nano-devices and some electronic and medical applications as well as the incorporation of materials and medicines.

Research is still ongoing in search of the latest methods that offer the lowest and least expensive environmental damage to continue. Similarly, research to discover new properties of copper nanoparticles and new products.

6. References

[1] U. Asim, N. Shahid and R. Naveed 2012 *World Scientific Publishing Company* **7**(5) 1230005
[2] J. A. Darr, J. Zhang, N. M. Makwana and X. Weng 2017 *Chem. Rev.* **117** 11125
[3] B. Khodashenas and H. R. Ghorbani 2014 *Korean J. Chem. Eng.* **31** 1105
[4] X. Y. Song, S. X. Sun, W. M. Zhang and Z. L. Yin 2004 *J. Colloid Interface Sci.* **273** 463
[5] X. D. Su, J. Z. Zhao, H. Bala, Y.C. Zhu, Y. Gao, S. S. Ma and Z. C.Wang 2007 *J. Phys. Chem. C.* **111** 14689
[6] C.W. Wu, B. P. Mosher and T. F. Zeng 2006 *J. Nanopart. Res.* **8** 965

[7] B. K. Park, S. Jeong, D. Kim, J. Moon, S. Lim and J. S. Kim 2007 *J. Colloid Interface Sci.* **311** 417
[8] A. A. Athawale, P. Prachi, M. Kumar and M. B. Majumdar 2005 *Mater. Chem. Phys.* **91** 507
[9] S. Panigrah, S. Kundu, S. K. Ghosh, S. Nath, S. Praharaj, B. Soumen and T. Pal 2006 *Polyhedron* **25** 1263
[10] A. P. Reverberi, M. Salerno, S. Lauciello and B. Fabiano 2016 *Materials* **9** 809
[11] P. Kanninen, C. Johans, J. Merta and K. Kontturi 2008 *J. Colloid Interface Sci.* **318**, 88
[12] M. P. Pileni 1993 *J Phy. Chem* **97** 6961
[13] H. Xu, B. W. Zeiger and K. S. Suslick, 2012 *Chem. Soc. Rev.* **26th**, DOI: 10.1039/c2cs35282f
[14] V. Sáez and T. J. Mason 2009 *Molecules* **14** 4284
[15] H. T. Zhu, C. Y. Zhang and Y. S. Yin 2004 *J. Cryst. Growth* **270** 722
[16] S. Giuffrida, L. L. Costanzo, G. Ventimiglia and C. Bongiorno 2008 *J. Nanopart Res.* **10** 1183
[17] S. Kapoor and T. Mukherjee 2003 *Chem. Phys. Lett.* **370** 83
[18] Q. B. Zhang and Y. X. Hua 2014 *Phys. Chem. Chem. Phys,* DOI: 10.1039/c4cp03041a
[19] S. L. Kadam 2016 *IJSER*, **7** 1067
[20] S. H. Yu 2001 *J. Ceram. Soc. Jpn.* **109** 65
[21] M. Rajamathi and R. Seshadri 2002 *Curr. Opin. Solid State Mater. Sci.* **6** 337
[22] A. Dinesh Karthik and K. Geetha 2014 *Int. J. Nano Dimens.* **5** 321
[23] N. Ahmad and N. Haram 2017 *Production and characterization of Cu nanoparticles by laser ablasion of solid Cu Target in Double Distilled Water*, www.ncp.edu.pk/docs/iss/talks/Group_I/G1_D1_%20Dr_Nisar_Ahmad.pdf
[24] S. Moniri, M. Ghoranneviss, M. R. Hantehzadeh and M. A. Asadabad 2017 *Bull. Mater. Sci.* **40** 37
[25] S. K. Yadav and V. Vasu 2016 *IJETST* **03** 3795
[26] W. Jiang and K. Yatsui 1998 *IEEE Trans. Plasma Sci.* **26** 1498
[27] K. Muraia, Y. Watanabea, Y. Saitoa, T. Nakayamaa, H. Suematsua, W. Jianga, K. Yatsuia, K.B. Shimb and K. Niiharaa 2007 *J. Ceram. Process Res.* **8** 114
[28] O. B. Nazarenko, A. P. Ilyin 2008 *Proceedings of the 3rd Environmental Physics Conference*, 135
[29] P.K. Dash and Y. Balto 2011 *Res. J. Nanosci. Nanotechnol.* **1** 25
[30] I. V. Chepkasova, Yu. Ya. Gafnera, S. L. Gafnera, and S. P. Bardahanov 2016 *Phys Metals Metallogr.* **117** 1003
[31] A. P. Zavjalov, K. V. Zobov, I. K. Chakin, V. V. Syzrantsev, and S. P. Bardakhanov 2014 *Nanotechnol. Rus.* **9** 660
[32] V. Bansal, D. Rautaray, A. Bharde, K. Ahire, A. Sanyal, A. Ahmad and M. Sastry 2005 *J. Mater. Chem.* **15** 2583
[33] R. Varshney, S. Bhadauria and M. S. Gaur 2012 *Nano Biomed. Eng.* **4** 99
[34] R. Varshney, S. Bhadauria, M. S. Gaur and R. Pasricha 2011 *Nano Biomed. Eng.* **3** 115
[35] K.V. Pavani, N. Srujana, G. Preethi and T. Swati 2013 *Lett. App. Nanobiosci.* **2** 110
[36] G. Shobha, V. Moses and S. Ananda 2014 *IJPSI* **3** 28
[37] H. J. Lee, G. Lee, N. R. Jang, J. M. Yun, J. Y. Song and B. S. Kim 2011 *Nanotech.* **1** 371
[38] A. Khan, A. Rashid, R. Younas and R. Chong 2016 *Int. Nano Lett.* **6** 21
[39] J. Xiong, Y. Wang, Q. Xue and X. Wu 2011 *Green Chem.* **13** 900
[40] L. M. Qi, J. M. Ma and J. L. Shen 1977 *J. Colloid Interface Sci.* **186** 498
[41] X. Y. Song, S. X. Sun, W. M. Zhang and Z. L. Yin 2004 *J. Colloid Interface Sci.* **273** 463
[42] J. Moghimi-Rad, F. Zabihi, I. Hadi, S. Ebrahimi, T. D. Isfahani and J. Sabbaghzadeh 2010 *J. Mater. Sci.* **45** 3804
[43] I. Haas, S. Shanmugam and A. Gedanken 2006 *J. Phys. Chem. B* **110** 16947
[44] S. L. Kadam 2016 *International Journal of Scientific & Engineering Research* **7** 1067
[45] TDS *Copper (Cu) nanopowder/nanoparticle (Cu, 99.9%, 40 nm, metal basis)* US. Research Nanomaterials, Inc.
[46] TDS *Copper (Cu) nanopowder/nanoparticle (Cu Nanoparticles with 5.2 wt% Cu_2O coated, 30 nm)* US. Research Nanomaterials, Inc.

[47] TDS *Copper (Cu) nanopowder/nanoparticle (Cu-Carbon Coated, 99.8%, 25 nm)* US. Research Nanomaterials, Inc.

Acknowledgements
Our participation in NNS 2017 is supported by the Bandung Institute of Science Technology (ITSB), Bekasi, Indonesia. We are grateful to the Ministry of Research, Technology and Higher Education of the Republic of Indonesia, and LPDP, for the doctoral scholarship program BUDI-DN 2016 (Grant number: 20161141011049).

Preliminary Study in Fabricating Fluorine-doped Tin Oxide by Using Spray Pyrolysis Methods

Darsikin[1], Z Arifin[2], Nasar[1] and S Saehana[1, a)]

[1)]Department of Physics Education, Tadulako University, Jln. Soekarno-Hatta km. 09 Kel. Tondo, Palu, Indonesia
[2)]Mechanical Engineering Departement, Sebelas Maret University, Jl. Ir. Sutami 36A Kentingan Surakarta 57126, Solo, Indonesia

[a)]Coresponding author: sahrulsaehana@gmail.com

Abstract. In this study, Fluorine-doped Tin Oxide (FTO) was fabricated by using spray pyrolysis methods for solar cells applications. The synthesis of FTO has been conducted under spray pyrolysis methods in which a precursor solution of $SnCl_2$: F having a concentration of 0.1, 03, and 0.9 M. The spray pyrolysis was performed using a nebulizer with a deposition time of 2, 3, 5 and 10 minutes. The glass substrate was heated at a temperature of 400°C prior to deposition. By Controlling concentration of $SnCl_2$: F, FTO glass was successfully produced. The lowest sheet resistance of the FTO glass was about 60 Ω/cm^2. X-Ray Diffraction and Scanning Electron Microscopy were employed to investigate the characteristic of FTO. Then, FTO was applied as an electrode in the dye sensitized solar cell (DSSC) electrode application and the photovoltaic effect was observed.

Keywords: FTO, spray method, dssc application.

1. Introduction

Until now, the application of a Transparent Conductive Oxide (TCO) is very wide [1-4]. An electronic device, such as solar cells [1-3], water splitting [4] and light emiting diodes [5], were developed by applying TCO. However, the price of these components is still very expensive. It is because the method used to create TCO requires complicated production costs [6-7].

One type of TCO is Flourine doped Tin Oxide (FTO). According to Ruso (2008), that FTO is an ideal candidate for applications requiring TCO due to its ability to adhere strongly to glass, resistance to physical abrasion, chemical stability, high optical visible transparency, and electrical conductivity [8]. FTO can be produced through spray pyrolysis methods [7-9], sol gel methods [10,11], chemical vapor deposition [12] and DC magnetron sputtering [13]. Most of the researcher using spray pyrolysis method because it is required low cost material and may produce at low temperature [8]. In this preliminary study, we made a Flourine doped Tin Oxide (FTO) by employing spray pyrolysis methods then it was applied as Dye Sensitized Solar Cells (DSSC) component. Futhermore, the performance of DSSC was performed under sun illumination.

2. Material and Methods

In this study, the materials used for the fabrication of the FTO glass were the solution of Tin(II) Chloride ($SnCl_2.2H_2O$, Merck, Germany), Ammonium Fluoride (NH_4F, Merck, Germany), and Ethanol (96%, Merck, Germany). The $SnCl_2$ and NH_4F materials of 1 gram each were fed into the ethanol solvent then stirred until homogeneous. Then, the spray pyrolysis method is used to deposit the solution on a glass substrate (1 cm x 1 cm). The concentration of solution and time deposition of nanoparticle are investigated to get the optimum resistance. After that, the sample was heated at 400°C for 30 minutes. A conductive transparent thin layer is then formed and conductivity measurements are made. Scanning Electron Microscopy was done to investigate the morphology of FTO film. Then, XRD diffraction is also employed to characterize the crystalization of FTO film.

3. Results and Discussion

Fabrication of FTO was performed by spray pyrolysis method (Fig. 1) [14]. It this method, nanoparticles were produced by spraying droplets from the solution onto the heated surface (Fig 1a). When a droplet contacts to a hot surface (in Fig. 1b), it will vaporize and the lagging oxide particles are then attached to the surface [9]. There are several parameters in affecting FTO resistance. It is solution concentration, time deposition, temperature and distance between nebulizer to the substrate [7].

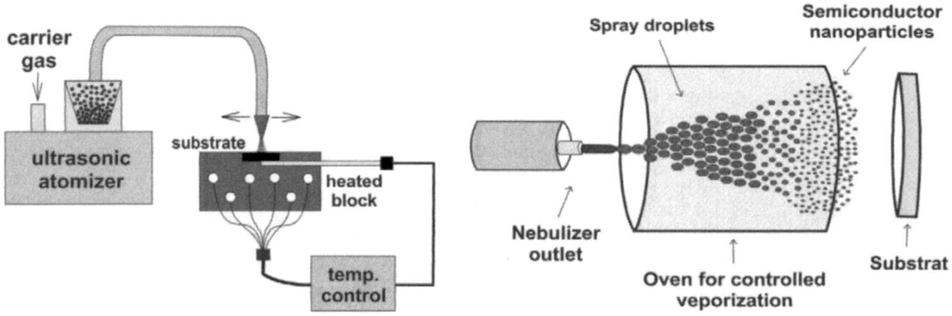

Figure 1. FTO fabrication process by using spray pyrolysis method: (a) Diagram of FTO deposition; (b) Process of the nanoparticles formation.

In this study, to obtain a good FTO resistance, optimization of two parameters was done by varying concentration of solution and deposition time. Effect of both parameters to sheet resistance was shown in Table 1.

Table 1. Value of FTO resistance with solution concentration and deposition time.

No	Concentration (M)	Time Deposition (minute)	Sheet Resistance (Ω)
1	0.9	2.0	9.66×10^6
2	0.9	5.0	1.07×10^6
3	0.9	10.0	1.62×10^6
4	0.3	5.0	65.6
5	0.3	10.0	99.3
6	0.3	10.0	324.6
7	0.1	3.0	1.27×10^5
8	0.1	10.0	3.25×10^5

From the results obtained, as shown in Table 1, it is known that both parameters (solution concentration and deposition time) affect the resulting FTO resistance. The resistance value of 65.6 to 324.6 Ω/cm^2 was found by combining the parameters. The good resistance was obtained at a concentration of 0.3 M and a 5-minute of deposition time. It was predicted that it was related to the loading level of nanoparticle formation on the substrate [15].

In order to know the crystalization and morphology of the FTO film, then the FTO sample was characterized by employing X-Ray Diffraction Spectroscopy (XRD) and Scanning Electron Microscopy (SEM).

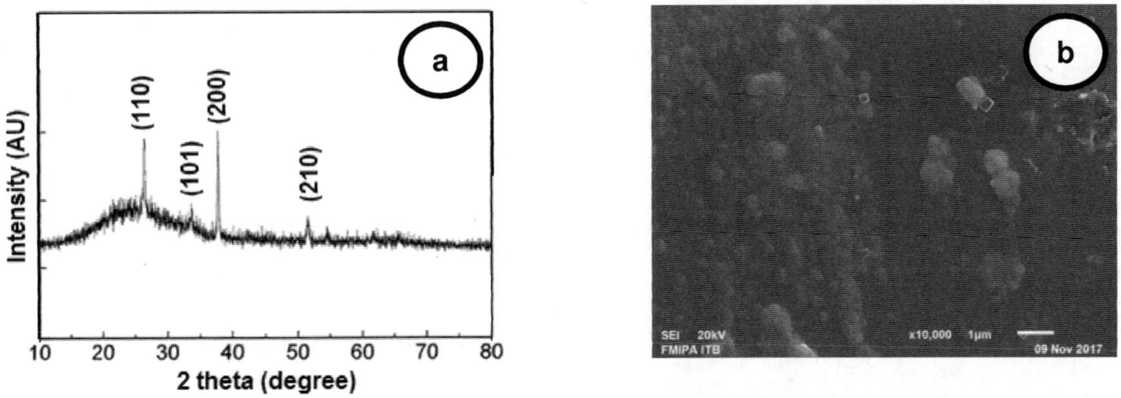

Figure 2. Characterization of FTO film: (a) XRD pattern of FTO films after annealing at temperature 400 °C; and (b) Scanning Electron Microscopy result.

Figure 2(a) shows an XRD pattern on the films after annealing at 400 °C for 30 min. The crystalline structure of the FTO films was tetragonal SnO_2. On the other hand, FTO nanoparticle grain can be observed at scanning electron microscopy result of FTO film in Figure 2(b). Furthermore, resulting FTO was applied as solar cells component. FTO was used as main and counter electrode of DSSC as shown in Figure 3(a). The performance of solar cells under the sun illumination can be seen in Figure 3(b).

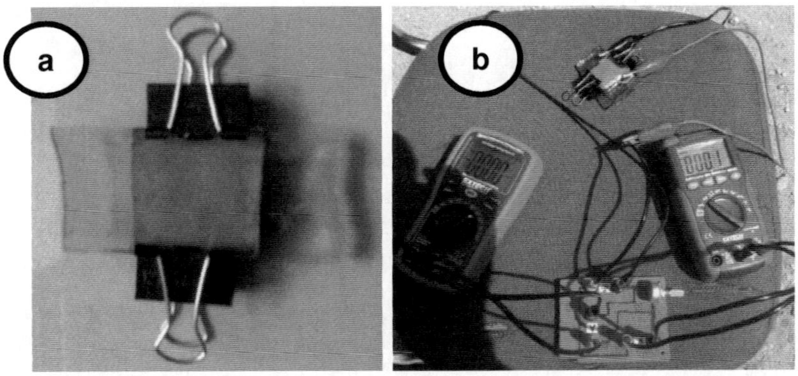

Figure 3. (a) DSSC solar cell with created FTO; (b) The measurement of DSSC under sun illumination.

The phenomenon of photovoltaics in DSSC was observed on 37.48 mW/cm^2 and IV characteristics were plotted in Fig. 4. Although the resulting performance of DSSC was still lack (in microwatt order) however FTO resulted have been able to be used as DSSC components as reported reference [16-18]. Furthermore, the researcher will make an efforts to obtain a better FTO in terms of conductivity, homogeneity, and transparency as suggested by the reference [16-18].

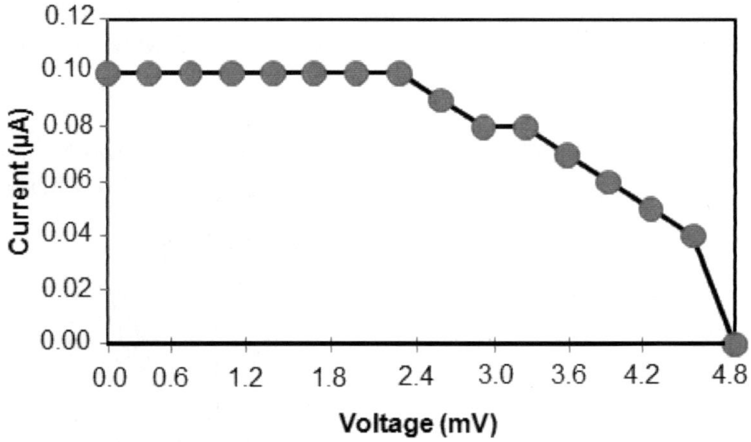

Figure 4. The IV Characteristics of DSSC by employing FTO produced.

4. Conclusion

In this study, the conductive thin layer of FTO with a good conductivity about 66 Ω/cm^2 has been successfully made using spray pyrolysis method. In the fabrication process, it was optimized concentration and time deposition. Crystalization and morphology of FTO were confirmed by employing XRD and SEM. The resulting FTO has also been tested as a DSSC component and the phenomenon of photovoltaics can be observed.

5. References

[1] Chiba, Y., et. al. 2006 *Japan. J. Appl. Phys.* **45**(25), pp. L638–L640.
[2] Dai, S., et. al. 2005 *J. Sol. Energy Mat. Sol. Cells* **85**, pp. 447–455.
[3] Gratzel, M., 2001 *J. Sol-Gel Sci. Tech.* **22** pp. 7–13.
[4] Noh E., et. al. 2013 *Sci. World J.* **2013** pp. 1-8.
[5] Chuang S-H, Tsung C-S, Chen C-H, Ou S-L, Horng R-H, Lin C-Y, and Wuu D-S 2015 *ACS Appl. Mater. Interfaces* **7** (4), pp 2546–2553.
[6] Chou, T.P., et. al. 2008 *J. Nanophotonics* **2** pp. 023511.
[7] Purwanto A, Widiyandari H, and Jumari A 2012 *Thin Solid Films* **520** (6), pp. 2092–2095.
[8] Russo B, and Cao G.Z 2008 *Appl. Phys. A* **90**, pp. 311–315.
[9] Suyitno, Arifin Z, Santoso A.A, Setyaji A.T, and Ubaidillah 2014 *Appl. Mech. Mat.* **575**, pp 689-695.
[10] Senthilkumar V, Vickraman P, and Ravikumar R 2010 *J. Sol-Gel Sci. Tech.* **53**(2) pp 316–321.
[11] Shi, X.H. and K.J. Xu 2017 *Mat. Sci. Semiconductor Process.* **58**, pp. 1–7.
[12] Chavarría-Castillo K.A et. al. 2016 *Mat. Research* **19** pp. 97-102.
[13] Banyamin Z.Y, Kelly PJ, West G, and Boardman J 2014 *Coatings* **4**, pp. 732-746.
[14] Montero, Chialvo, M.R.G and Chialvo, A.C 2009 J. Mater. Chem. **19** 3276–3280.
[15] Supriyono, Surahmana H, Krisnandia Y.K, and Gunlazuardia 2015 *J. Procedia Environ. Sci.* **28** pp. 242–251
[16] Saehana S, Darsikin, Yuliza E, Khairurrijal and Abdulah M 2012 *J. Sol. Energy Eng.* **136**, pp. 044504-1.
[17] Saehana S, Muslimin, and Abdulah M 2014 *J. Renew. Sustain. Energy.* **6**, pp. 023109.
[18] Saehana S, Arifin A, Khairurrijal, Abdulah M 2012 *J. Appl. Phys.* **111**, 123109.

Acknowledgemet

This work was partially supported by Teacher and Training Faculty of Tadulako University Research Grand and Riset Terapan Unggulan Perguruan Tinggi Decentralization Research Grant from the Research, Technology and Higher Education of the Ministry of the Republic of Indonesia in the fiscal years of 2017. The researcher also gives appreciation to 1) Scanning Electron Microscopy Laboratory in Faculty of Mathematic and Natural Sciences of ITB, and 2) Functional Laboratory of Faculty of Mathematic and Natural Science of Makassar State University.

IOP Conf. Series: Materials Science and Engineering **395** (2018) 012016 doi:10.1088/1757-899X/395/1/012016

LabVIEW Simulation for Electroplating Process of Plated-Through Hole (PTH) in Multilayer Printed Circuit Board (PCB)

M Arifin[1,2*], A B K Atmaja[1], and V Octowinandi[1,2]

[1]Department of Electrical Engineering, Politeknik Negeri Batam, 29461 Batam, Indonesia
[2]Teaching Factory Manufacturing of Electronics, Politeknik Negeri Batam, 29461 Batam, Indonesia

*E-mail: arifin@polibatam.ac.id

Abstract. Plated-through hole (PTH) is essential part which it connects one layer to other layers in multilayer printed circuit board (PCB). PTH is commonly made by electroplating process with copper as based conductive material. This research objective to simulate the electroplating process in which mass of copper generated using LabVIEW which it is used together with Faraday's law of electrolysis. The simulation using LabVIEW software and experiment are applied in this study. The simulation and the calculation results of copper mass on the PCB and PTH surfaces were displayed in LabVIEW interface. On the other hand, the experiment is also conducted to form PTH and determine the mass of copper experimentally on the PCB and PTH surfaces. The experiment results showed that PTH was successfully formed in multilayer PCB with electroplating process. The results of copper mass adhered in multilayer PCB in which they were obtained from LabVIEW simulation or calculation are not much different with the copper masses obtained experimentally. The copper masses obtained from experiment tend smaller than the calculation.

1. Introduction

Electroplating is one of the hardest surface adjustment technologies for pre-treatment product. Besides used in electrochemical impedance of lithium battery [1], electrochemical reaction and electric current are applied in this electroplating for making metal coatings. The coating materials frequently used in electroplating are silver, gold, tin, zinc, cadmium, copper, nickel, chromium, and so on [2]. Electroplating is widely used in electronics, aerospace, automotive [3], microelectronic mechanic system (MEMS) technology [4,5,6] and engineering industries. This electroplating has many advantages such as corrosion prevention, aesthetic finishes, coating component [3], lower resistivity, better anti-electromigration ability, and higher thermal exchange rate [7].

The electroplating results that consist of surface smooth, copper grain, and resistivity are affected by many parameters such as electric field distribution, electric current density, temperature and stirring style [8,9]. In copper electroplating, the most dominant factor is the current density because the electrodepositing speed and the grain size are decided by the current density [10]. The copper electroplating is commonly used in printed circuit board (PCB) manufacturing [11,12]. The double layer and multilayer PCB need the electroplating process to make the conductive surface in plated-through hole (PTH) [12].

In order to learn more about the electroplating process in multilayer PCB, the simulation of electroplating process is needed. In this study, LabVIEW simulation is utilized to learn the electroplating process in multilayer PCB. LabVIEW is applied together with Faraday's law of electrolysis.

2. LabVIEW Simulation

LabVIEW programs are also called virtual instruments. A LabVIEW virtual instrument consists of a front panel, a block diagram, and a connector pane. A front panel is considered as a user interface. In front panel, the user can insert the inputs (sliders, knobs, values entered from the keyboard, push buttons, selector switches) and the outputs (indicators, LEDs, graphs, strip charts, sounds) [13].

A block diagram is considered as a LabVIEW program. Block diagrams integrate terminals, subVIs, functions, constants, structures, and wires, which transfer data among other block diagram objects. Furthermore, a connector panel is a set of terminals on the icon that corresponds to the controls and indicators of a VI, similar to the parameter list of a function call in text-based programming languages [13]. In this research, LabVIEW is applied together with Faraday's law of electrolysis as in equation (1) [3].

$$m = \frac{A}{n} \frac{I\,t}{96500} \tag{1}$$

Where *m, A, n, I, t*, and 96500 are mass of deposited material (gram), atomic weight of deposited material, valency of deposited material, applied current (Ampere), time (s), and Faraday's constant (Coloumb/gram), respectively [3].

3. Experimental procedure

The copper mass adhering on printed circuit board (PCB) and plated-through hole (PTH) surface is also determined with the experiment. The experiment was conducted through some steps such as desiging PTH in PCB, preparing material, pressing the FR-4 material, drilling, brushing, cleaning, immersing in the conductive carbon solution, drying, electroplating process, cleaning, drying, and measuring the copper mass.

The prepared material was FR-4 PCB material, distilled water, conductive carbon solution (black hole solution), and copper (II) sulfate pentahydrate ($CuSO_4.5H_2O$, meltex). The first step made a PCB stack from three FR-4 PCB materials using glue as adhesive and pressing them to make PCB with four layers. The pressing process was conducted at 12 bar with temperature 170 ^oC for 30 minutes. The second step was the drilling process which it works following the PCB design. The PCB design was created using Altium (PCB design software). The third step was brushing and cleaning. The distilled water was used during brushing and cleaning. The fourth step is immersing the PCB surface in the conductive carbon solution (black hole solution) for 5 minutes and drying at 80 ^oC for 10 minutes. The fifth step is the electroplating process. Before the electroplating process is running, the multilayer PCB was weighed with the scale (Kern FCB bench scale). The electroplating process was conducted for three PCB, but the applied current were different for each PCB in which the applied currents were 15 A, 20 A, and 25 A. Each electroplating process needed 45 minutes. Cleaning and drying of PCB were done after the electroplating process was finished. The last step is weighing the copper mass adhering on the PCB and PTH surfaces.

4. Results and discussion

LabVIEW software is applied to simulate how to work electrochemically the electroplating process. It simulates the movement of copper ion in copper(II) sulfate ($CuSO_4$) solution when the electrodes are given the current. On the other hand, LabVIEW also calculates the copper mass that adhering on the PCB and PTH surfaces. The equation (1) is applied in this calculation.

LabVIEW interface displays the table of input data (current, time, copper mass), the simulation of copper ionic movement, and the graph. The graph exhibits the relation between the applied current and

IOP Conf. Series: Materials Science and Engineering **395** (2018) 012016 doi:10.1088/1757-899X/395/1/012016

the copper mass. The front panel also displays other parameters such as atomic weight of deposited material, valency of deposited material, and button for 'play' and 'stop' as shown in figure 1.

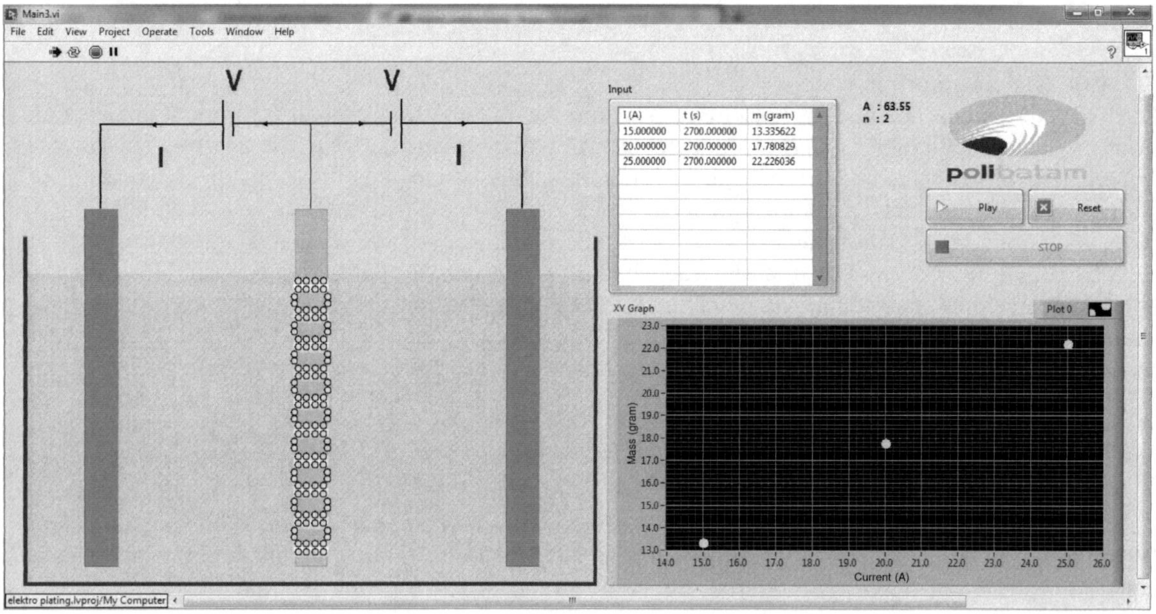

Figure 1. Interface of LabVIEW virtual instrument for electroplating simulation.

LabVIEW software utilizes the graphical program language placed in the diagram block. The program receives the input data from the front panel such as parameter data, 'play' button, 'stop' button, etc. All program used in electroplating simulation and calculation of copper mass are placed in the diagram block as shown in figure 2 and subprogram for the calculation of copper mass during simulation of electroplating process as shown in figure 3.

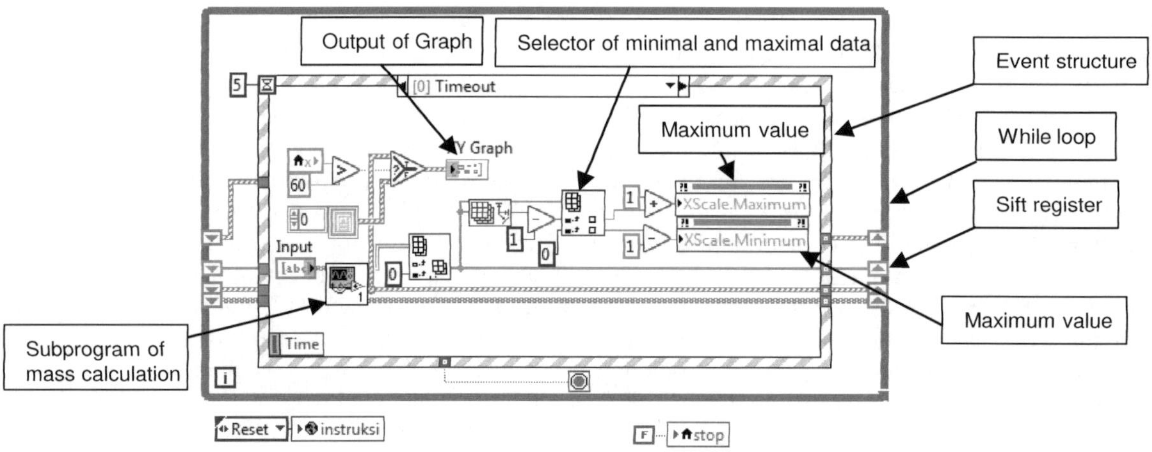

Figure 2. Block diagram of LabVIEW virtual instrument.

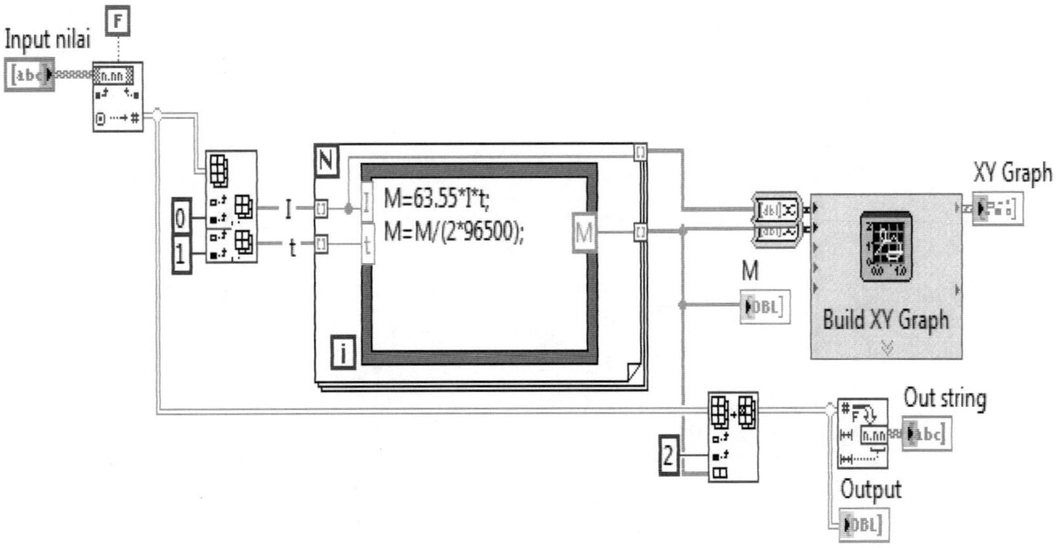

Figure 3. Subprogram for the calculation of copper mass during simulation of electroplating process.

The calculation results of copper atomic masses adhering on the PCB and PTH surfaces using the equation (1) are 13.3 gram, 17.8 gram, and 22.2 gram for the applied current 15 A, 20 A, and 25 A, respectively. Besides calculation, the copper masses are also obtained experimentally. The copper masses are obtained from difference between the PCB mass before and after the electroplating process. Multilayer PCB before and after the electroplating process are shown in figure 4.

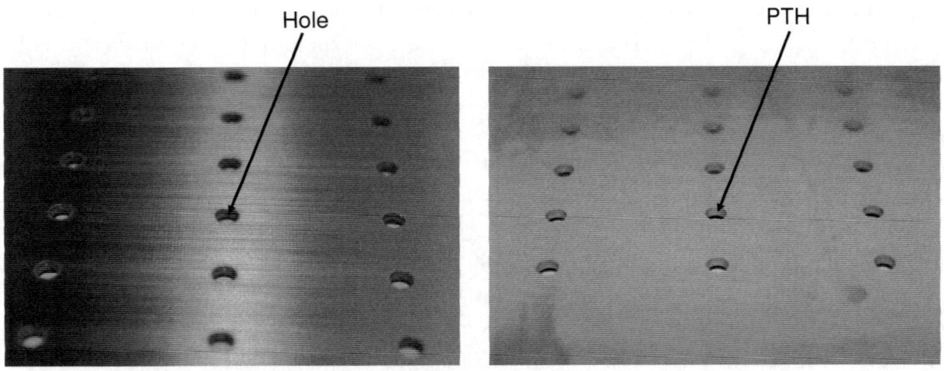

(a) (b)

Figure 4. Multilayer PCB (a) before and (b) after the electroplating process.

The calculation results are slightly different with the experiment results. The experiment results showed that the copper masses on the PCB and PTH surfaces are 13 gram, 16 gram, and 20 gram for the applied current 15 A, 20 A, and 25 A, respectively. There are slightly different between calculation results and experiment results for copper mass as shown in figure 5. The different results for the copper masses are possibly caused by impurity of $CuSO_4$ solution and the current difference in amperemeter and the real current.

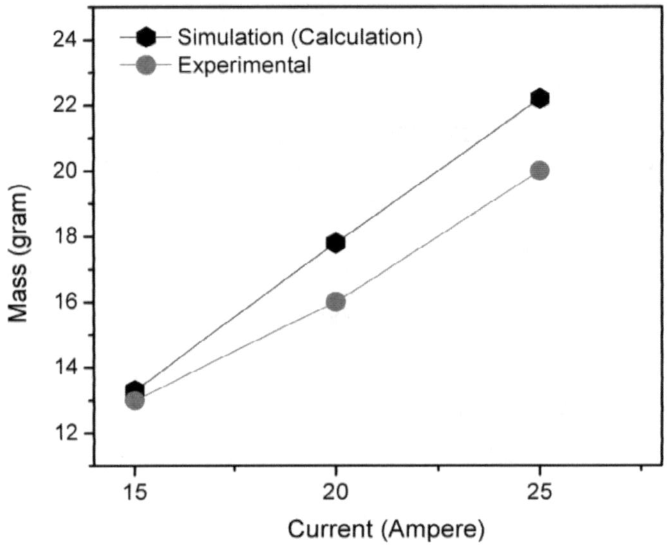

Figure 5. The copper masses obtained from calculation and experiment.

5. Conclusion

In this study, the simulation of electroplating process in multilayer printed circuit board (PCB) was successfully created by LabVIEW software. The calculation results for copper mass adhering on the PTH and PCB surfaces are 13.3 g, 17.8 g, and 22.2 g with the applied current 15 A, 20 A, and 25 A, respectively. Furthermore, the experiment results exhibit that the copper masses are 13 g, 16 g, and 20 g for the applied current 15 A, 20 A, and 25 A, respectively. The results of copper mass adhered in multilayer PCB obtained from LabVIEW simulation or calculation is not much different with the mass of copper obtained experimentally.

6. References

[1] Arifin M, Rus Y B, Aimon A H, Iskandar F, Winata T, Abdullah M and Khairurrijal K 2017 *Mater. Res. Express* **4** 034005

[2] Selhi A and Nikhil S 2014 *Proc. Int. and All India Manufacturing Technology, Design and Research Conf.* (Assam: Indian Institute of Technology Guwahti) p 439-1

[3] Masuku E S, Mileham A R, Hardisty H, Bramley A N, Johal C and Detassis P 2002 *CIRP Ann Manuf Technol* **51** 168

[4] Oh S H, Lee K-C, Chun J and Lee M K S S 2001 *J. Micromech. Microeng.* **11** 221

[5] Sadler D J, Gupta S and Ahn C H 2001 *IEEE Trans. Magn.* **37** 2897

[6] Lin M-T, Tong C-J and Chiang C-H 2006 *DTIP of MEMS and MOEMS Stresa*

[7] Chen Y, Li X, Chen S Q and Zhang H 2008 *ASME Proceedings* 619

[8] Hu C-C and Wu C-M 2003 *Surf. Coat. Technol.* **176** 75

[9] Seah C H, Mridha S and Chan L H 1999 *J. Mater. Process. Technol.* **89** 432

[10] The W H, Koh L T, Chen S M, Xie J, Li C Y and Foo P D 2001 *Microelectronics J.* **32** 597

[11] Khandpur R S 2006 *Printed Circuit Board: Design, Fabrication, and Assembly* (New Delhi: McGraw-Hill) p 313

[12] Lin G-Y, Yan J-J, Yen M-Y, Dow W-P and Huang S-M 2013 *J. Electrochem. Soc* **160** D3028

[13] Boubaker O 2011 *Proc. Int. Conf. on Frontiers in Education: Computer Science and Computer Engineering* (Tunis: National Institute of Applied Sciences and Technology)

IOP Conf. Series: Materials Science and Engineering **395** (2018) 012016 doi:10.1088/1757-899X/395/1/012016

Acknowledgments
The authors thank to the Young Lecturer Research Fund Program 2017 from Politeknik Negeri Batam. The authors also thank to Teaching Factory Manufacturing of Electronics (TFME) from Politeknik Negeri Batam that it allowed to use all equipment especially in PCB manufacturing equipment.

IOP Conf. Series: Materials Science and Engineering **395** (2018) 012017 doi:10.1088/1757-899X/395/1/012017

Preparation of Stable Colloidal Silica with Controlled Size Nano Spheres from Sodium Silicate Solution

L Qomariyah, F N Sasmita, H R Novaldi, W Widiyastuti, and S Winardi*

Department of Chemical Engineering, Institut Teknologi Sepuluh Nopember, Kampus ITS Sukolilo, Surabaya 60111, Indonesia

*swinardi@chem-eng.its.ac.id

Abstract. The formation process of stable colloidal silica solution was discussed in this article. The affecting parameter including pH, zeta potential, and colloidal concentration was investigated to elucidate the stability of colloid. Sodium silicate was used as the silica source. The colloidal silica prepared via sol-gel method at the base condition. Sodium ion of sodium silicate was removed by ion-exchanging to obtain the active silicic acid which was titrated with potassium hydroxide (KOH) solution to form the colloidal silica. Size distribution and solution zeta potential were measured by Particle Size Distribution (PSA). In this experiment, the results showed that the higher solution pH (affect the decreasing of zeta potential value) and higher colloidal concentration (affect the increasing of zeta potential value) play important role in the stability of colloidal silica and controlling the particle growth to form the larger particle size. The colloidal silica formed at this condition was agglomeration free for a long period that indicated the colloids were stable.

Keywords: Colloidal silica, Sodium silicate, Stable colloid

1. Introduction

The formation process of colloidal silica via sol-gel method is of interest due to the fact that colloidal silica can be used in various fields. The colloidal silica has a series of good physicochemical properties, e.g. strong adsorb ability, large surface area and good binding properties [1]. The colloidal silica has potential application in casting, stiffener in hard coating reagent, and textile [2][3][4]. Recently, it has been applied to make paper using micro particles [5]. In making paper, a stable dispersion is very important to inhibit particle's coalescing into larger agglomerates. Many commercial products already made using colloidal silica but used tetra esters as the raw materials [6][7][8]. However, this material was very expensive and toxic. It is necessary to use an abundant resource of silica such as sodium silicate solution or well known as water glass.

Colloidal silica has been synthesized by various method including sol-gel method [9][10][11], hydrothermal reaction [12][13], and chemical vapor deposition (CVD) [14]. The sol-gel method widely used to produce pure silica particles due to its ability to control the particle size and morphology through systematic monitoring of reaction parameters in ambient temperature [15].

The process used for making colloidal silica from sodium silicate using sol-gel method included ion-exchanging method [16]. This method can easily control the particle size and distribution in colloidal silica. Control of silica size in colloidal precursor was important due to it would influenced the bulk particle produced and the colloid stability. It enhanced by provide an electric charge on the surface of dispersed particles. In colloidal system, high zeta potential value will provide the stability of the solution to repulsive the aggregation [17]. The like charge on the particles will prevent the agglomeration between particles. Many experiments conducted using sodium silicate as raw material, but only few kinds of literatures had discussed the controlling silica size in the colloidal solution by giving an electric charge on surface of dispersed particles. Tsai [3] has been studied the process parameters of the formation of colloidal silica via sodium silicate. The formation temperature, titration rate and alkali concentration affected the surface growth of silica particle. However, the colloidal silica particle formed in this method was unstable at high formation temperature and titration rate of silica seed.

In present study, we report the synthesized of colloidal silica via sodium silicate at ambient temperature. This method has low heating consumption and formed stable colloidal silica. The colloidal silica was unstable at high formation temperature. So, it is important to study the colloidal silica formation at ambient temperature. Otherwise, colloidal silica pH value will give different surface particle charge as the present of different amount of potassium hydroxide concentration. Therefore, we investigated the effect of solution pH and solution concentration on the colloid stability and particle growth as the present of surface charge on the silica particles.

2. Experimental

Sodium silicate ($Na_2O.nSiO_2.mH_2O$, 8% Na_2O, 27% SiO_2, Merck) and potassium hydroxide (KOH, 85%, Merck) were used as received without further purification. Colloidal silica particles in the case of basic silica sol were prepared based on a method established by Liu [1]. Water glass diluted in distilled water to obtain 0.1, 0.2, and 0.3 N colloidal silica solution at ambient temperature. Subsequently, the diluted water glass was passed through a cation exchange resin (Amberlite^{TM} IR120, Dow Chemical) to form active silicic acid solution. The sodium ion of the dilute sodium silicate solution was substituted by hydrogen ion on the exchange sites of cation resin.

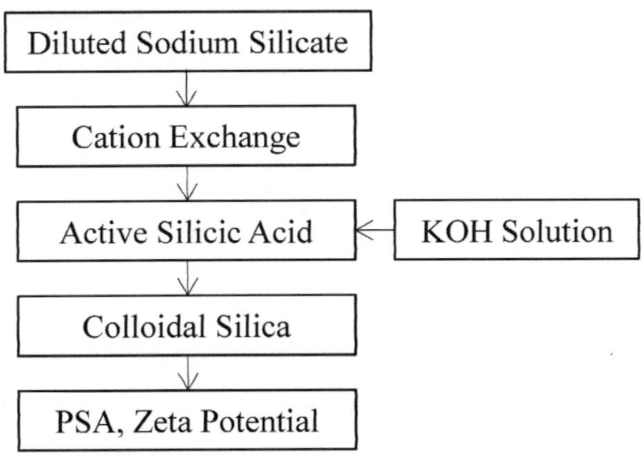

Figure 1. The experimental flowchart of formation colloidal silica.

Potassium hydroxide solution (0.1 M) was added drop-by-drop while the solution was stirred to adjust the solution pH from 8-11. At this condition, the polymerization of silicic acid leads to the formation of silica particles dispersed in water to give hydrosols. The precursors were used to study the effect of solution pH and concentration on the morphology of silica particles during the drying process. The experimental flowchart of formation colloidal silica had shown in figure 1. The stability of colloidal precursor very important in its application. The formed colloidal silica solutions were characterized by particle size analyzer and zeta potential (Malvern Zetasizer) to measure the particle size distribution and particle size.

IOP Conf. Series: Materials Science and Engineering **395** (2018) 012017 doi:10.1088/1757-899X/395/1/012017

3. Result and Discussion

Figure 2 shows the silica particle size distribution in colloidal solution in different solution concentration at pH 8. It can be observed that the size of silica particle in colloidal precursor increased as the solution increased. The increasing of particle caused by the aggregation of seeds when the solution pH suddenly increased during the addition of KOH. The amount of silica particle in solution increased as the result of the increasing solution concentration from 0.1 N to 0.3 N. The homogeneous nucleation of active silicic acid in precursor solution, have been occurred first before the addition of KOH solution. The active silicic acid going through nucleation and polymerization process. Silicic acid undergoes the polymerization from monomer become dimer and finally silica particle. The particle growth from nanometer size to micrometer size [18]. The schematic of silica formation from sodium silicate with salt absent present in figure 3. The effect of solution pH on the particle size also described. The solution pH arranged at basic condition from 8 to 11. The particle size tends to increase from 3.146, 3.448 and 4.066 nm when the solution pH increased from 8 to 11 respectively. However, the particle size decreased become 3.55 nm at pH 11. At pH 11, all solid phases of silica dissolve in water, this resulted slow precipitation which causes the particles formed in droplets become small. This happens because of the hydroxyl ions that catalyze the dilution process of small particles that limit the formation of larger particles [19]. The particle formation of colloidal silica particle affected by the addition of potassium hydroxide solution.

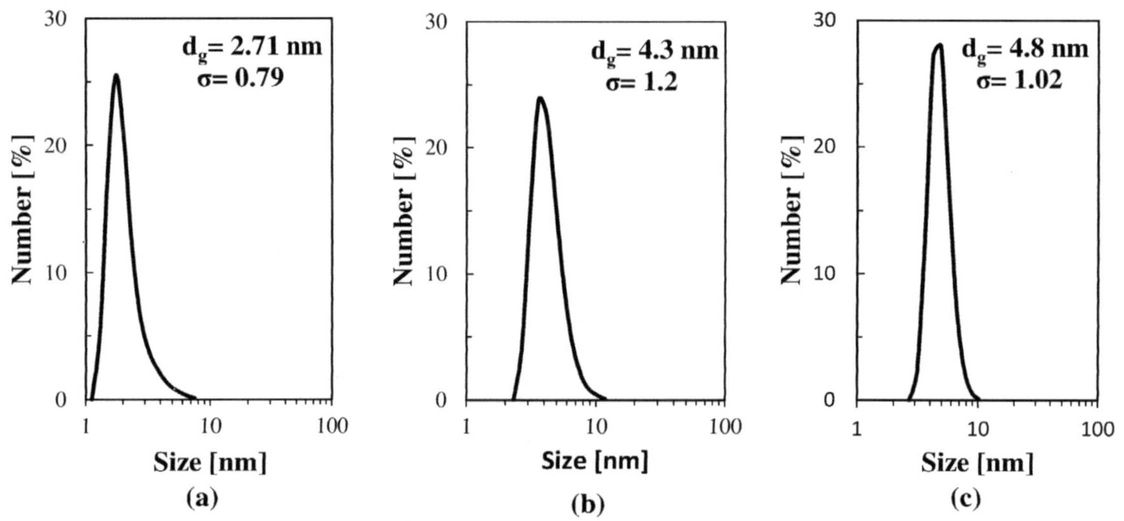

Figure 2. (a) The effect of solution concentration (a) 0.1 M; (b) 0.2 M; (c) 0.3 M to the particle size at pH 8.

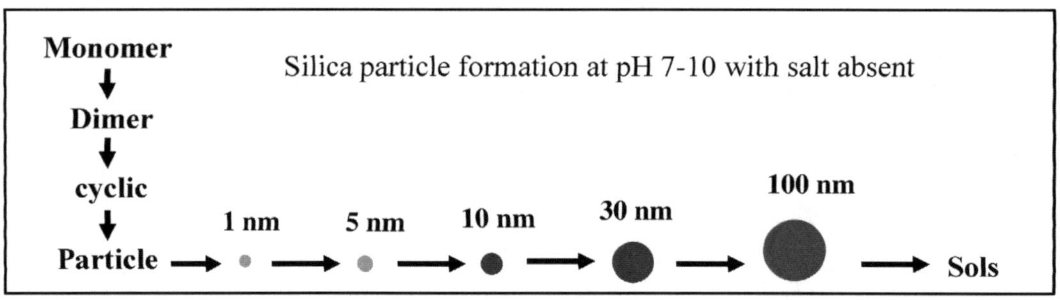

Figure 3. Silica particle formation at pH 7-10 with salt absent.

IOP Conf. Series: Materials Science and Engineering **395** (2018) 012017 doi:10.1088/1757-899X/395/1/012017

To investigate the effect of solution concentration to the stability of colloidal silica solution, zeta potential (ζ) measurement was conducted by Malvern Zetasizer instrument. Figure 4 (a) shows the effect of solution pH to the zeta potential value. Zeta potential value related to the particle surface charge, a property that all materials possesses, or acquire, when suspended in fluid. A high zeta potential value will provide the stability of the solution to repulsive the aggregation. The like charge on the particles will prevent the agglomeration between particles and thus maintaining the Brownian motion [20]. Colloidal stability achieves when the absolute value of $\zeta > +10$ (positive charge) and $\zeta > -10$ (negative charge). Based on Fig. 4 (a), the absolute zeta potential value higher than -10 indicated that the colloidal silica was stable. The value decreased as the solution pH increased. At the higher pH value, the extra electrolyte in the solution compressed the thickness of electrical double layer on the colloidal silica which made the zeta potential value decreased. This condition made the particle growth more easily. From this result, it is indicated that the solution pH can control the stability of colloidal silica and the particle size.

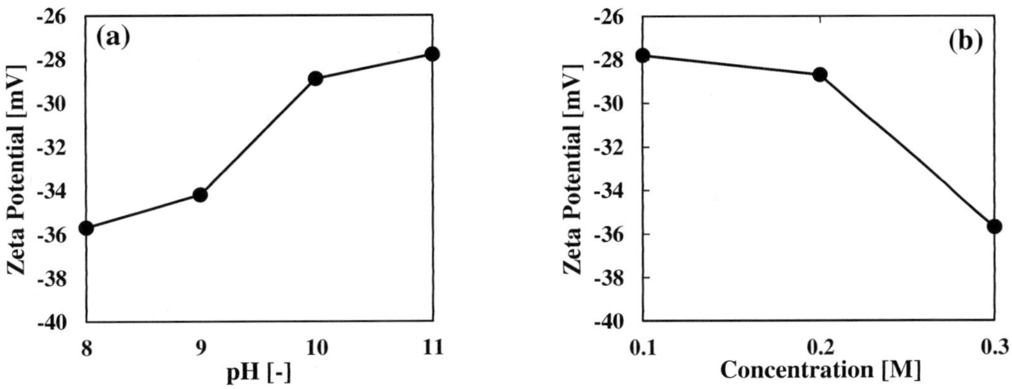

Figure 4. (a) The effect of solution pH to zeta potential; (b) The effect of solution concentration to zeta potential.

Figure 4 (b) shows the effect of solution concentration to the zeta potential value. The thickness of electrical double layer depends upon the concentration of electrolyte on the surface charge. As the concentration increased the amount of active silicic acid also increased, but the amount of electrolyte remains constant. This result indicated that the absolute zeta potential increase as the electrolyte concentration that compress the electrical double layer decrease. The absolute zeta potential value higher than -10 indicated that the colloidal silica was stable. Different with the effect of solution pH, the particles growth in concentration effect due to the diffusion mechanism between particles has a strong effect. From this result, it is indicated that the solution concentration can control the stability of colloidal silica and the particle size.

Figure 5 (left) shows the photographs of colloidal silica after being storage for several times. The colloidal silica solutions have transparent appearance. The colloidal solution remains transparent for more than ten days. The increasing of solution concentration from 0.1 N to 0.3 N did not affecting the stability of solution. The stability of colloidal silica solution affected by zeta potential value. Higher zeta potential of silica particle value maintains the particle interaction to repulsive each other. This result indicated that the colloidal silica was stable, and no agglomeration occurs for a long times storage. The longer storage time for 30 days, the same colloidal solution remains transparent. It can be seen in Fig. 5 (right). This condition also happens in another pH condition (9, 10 and 11) in the same solution concentration range (not shown here). The appearance of colloidal solutions is transparent after the long storage time more than 30 days.

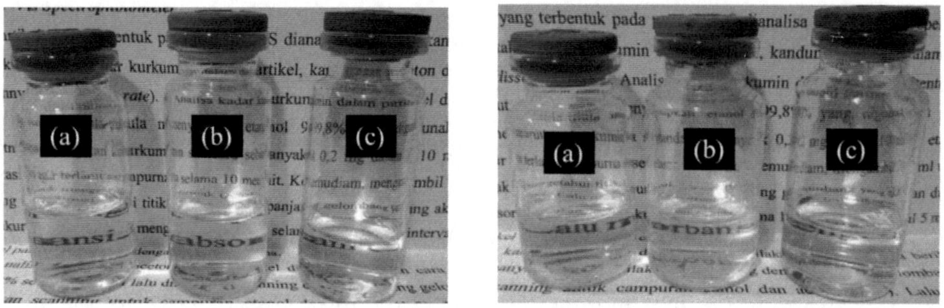

Figure 5. Photographs of colloidal silica after being storage for 10 days (left) and 30 days (Right) at pH 8 (a) 0.1 N; (b) 0.2 N; (c) 0.3 N.

4. Conclusion

The process parameter of the formation of colloidal silica using sodium silicate via sol-gel method had been investigated. The results showed that the process parameters such as solution pH and solution concentration affected the colloidal stability and particle size distribution. The mean particle size increased with increasing solution pH and increasing solution concentration. The value of zeta potential particle above ±10. The colloidal silica produced at this condition was agglomeration free for a long period, indicated that the colloidal silica was stable.

5. Reference

[1] Cheng Liu H, Xi Wang J, Mao Y and San Chen R 1993 The preparation and growth of colloidal particles of concentrated silica sols *Colloids Surfaces A Physicochem. Eng. Asp.* **74** 7–13

[2] Lee Y K, Rock Yoon Y and Rhee H K 2000 Preparation of colloidal silica using peptization method *Colloids Surfaces A Physicochem. Eng. Asp.* **173** 109–16

[3] Tsai M 2004 The study of formation colloidal silica via sodium silicate *Mater. Sci. Eng. B* **106** 52–5

[4] Tsai M S, Huang P Y and Wu W C 2005 The study of formation process of colloidal silica *Mater. Res. Bull.* **40** 1609–16

[5] Towle G A 1976 United States Patent **1** 0–3

[6] Stöber W, Fink A and Bohn E 1968 Controlled growth of monodisperse silica spheres in the micron size range *J. Colloid Interface Sci.* **26** 62–9

[7] Chen S-L, Dong P, Yang G-H and Yang J-J 1996 Characteristic Aspects of Formation of New Particles during the Growth of Monosize Silica Seeds *J. Colloid Interface Sci.* **180** 237–41

[8] Coenen S and Dekruif C G 1988 Synthesis and Growth of Colloidal Silica Particles *J. Colloid Interface Sci.* **124** 104–10

[9] Luo X, Dong J, Zhang L, Du J, Wang H and Gao W 2017 Preparation of silica micro spheres via a semibatch sol–gel method *J. Sol-Gel Sci. Technol.* **81** 669–77

[10] Bogush G H, Tracy M A and Zukoski IV C F 1988 Preparation of monodisperse silica particles: Control of size and mass fraction *J. Non. Cryst. Solids* **104** 95–106

[11] Giesche H 1994 Synthesis of monodispersed silica powders ii: controlled growth reactions and continuous production processes *J. Eur. Ceram. Soc.* **14** 205–14

[12] Park D H, Nishiyama N, Egashira Y and Ueyama K 2001 Enhancement of hydrothermal stability and hydrophobicity of a silica MCM-48 membrane by silylation *Ind. Eng. Chem. Res.* **40** 6105–10

[13] Song H, Rioux R M, Hoefelmeyer J D, Komor R, Niesz K, Grass M, Yang P and Somorjai G A 2006 Hydrothermal growth of mesoporous SBA-15 silica in the presence of PVP-stabilized Pt nanoparticles: Synthesis, characterization, and catalytic properties *J. Am. Chem. Soc.* **128** 3027–37

[14] Method V D, Okada S, Matsumura H, Displays O L D, Heya A, Niki T, Temperatures L,

IOP Conf. Series: Materials Science and Engineering **395** (2018) 012017 doi:10.1088/1757-899X/395/1/012017

Takano M, Niki T, Applications R and Tehrani F S 1998 Related content Formation of Silicon-Based Thin Films Prepared by Catalytic Chemical Vapor Deposition (Cat-CVD) Method

[15] Rahman I A and Padavettan V 2012 Synthesis of Silica nanoparticles by Sol-Gel: Size-dependent properties, surface modification, and applications in silica-polymer nanocompositesa review *J. Nanomater.* **2012**

[16] Tanev P T and Pinnavaia T J 1996 Mesoporous Silica Molecular Sieves Prepared by Ionic and Neutral Surfactant Templating: A Comparison of Physical Properties *Chem. Mater.* **8** 2068–79

[17] Ui S and Lim S 2009 Control of the size and morphology of nano-size silica particles using a sodium silicate solution *J. cermic Process. Res.* **10** 4–9

[18] Anon 1979 Iler, R. K., 1979. "The Chemistry of Silica (Solubility, Polymerization, Colloid and Surface Properties, and Biochemistry)". John Wiley&Sons, New York

[19] Setyawan H, Yuwana M and Winardi S 2007 Effect of Solution pH on the Morphology of Silica Particles Prepared by the Spray Drying of Sodium Silicate Solution

[20] Particle Sciences 2012 An Overview of the Zeta Potential *Part. Sci.* **2** 1–4

Acknowledgement

The authors are grateful for financial support provided by a grant from Directorate General of Higher Education, Ministry of Education and Culture of Indonesia through the scholarship Program Magister Doktor Sarjana Unggul (PMDSU) Batch II 2015.

IOP Conf. Series: Materials Science and Engineering **395** (2018) 012018 doi:10.1088/1757-899X/395/1/012018

The Influence of Operating Parameters on Membrane Performance in Used Lube Oil Processing

D Ariono[1], S Widodo[1,2], K Khoiruddin[1], A K Wardani[1], I G Wenten[1],*

[1] Department of Chemical Engineering, Institut Teknologi Bandung, Jl. Ganesha 10, Bandung 40132, Indonesia.

[2] PPPTMGB LEMIGAS, Balitbang KESDM, Jl. Ciledug Raya, Kav. 109, Cipulir, Kebayoran Lama, Jakarta, 12230, Indonesia.

*Corresponding author: igw@che.itb.ac.id

Abstract. Membrane process is an attractive technology for used lube oil treatment due to the advantages, i.e. relatively low energy consumption, intensive process, higher separation efficiency, and the improvement of final product quality. The performances of membrane processes depend on several factors such as feed characteristic, membrane properties, and operating parameter. In this work, the influences of solute concentration and applied pressure on the performance of hollow fiber polypropylene membrane during used lube oil filtration are investigated. The osmotic pressure model is used to simulate the influence of operating parameter on membrane performances and it results in good agreement with experimental data. By using the osmotic pressure model, the intrinsic properties of the membrane, namely membrane resistance (R_m) and mass transfer coefficient (k) are $5x10^6$ atm.s/m and $5x10^{-9}$ m/s, respectively. Generally, the flux increases with applied pressure and decreases with solute concentration. However, when the osmotic pressure builds up, the flux decreases gradually reaching a limiting value. At this condition, increasing an applied pressure gives an insignificant effect and the flux is in the limiting flux region. When the permeate flux reaches a limiting flux, the flux will be 5% of or less than the pure solvent permeability.

1. Introduction

Membrane-based process is an attractive technology because of their features such as lower energy consumption, more intensive process, higher separation efficiency, and higher product quality [1-5]. With those advantages, membrane has been widely used in various applications. Membrane applications include food processing and biotechnology, chemical processing [6-10], energy conversion [11], gas separation [12, 13], water treatment [14-22], and wastewater treatment [23-26]. In addition, membrane can also be used to treat and recover valuable component such as base oils from used lube oils, which will result in a waste minimization and an environmental protection.

Generally, used lube oils contain sludge, soot, carbonaceous particles, unburned fuel, tar, polycyclic aromatic hydrocarbons (PAHs), chlorinated paraffin and poly-chlorinated biphenyls (PCBs), metal-additives and wear metal of engines parts [27-29]. Several membranes were used and studied for used oils processing, e.g. polyethersulphone (PES), polyvinylidenefluoride (PVDF), polyacrylonitrile (PAN) [27], polyimide [30], and ceramic membranes [31-34]. Results showed that the membrane exhibited a better removal of ash content (55–75%) than the conventional method [35].

The performances of membrane processes depend on various factors including the characteristic of feed, the properties of the membrane, and the operating parameters [36]. For example, in the used oil filtration, the solute concentration significantly affect the performance of a membrane [37]. In this work, the influence of solute concentration and applied pressure on the performance of hollow fiber polypropylene membrane during used lube oil filtration is investigated. A simulation is conducted by the osmotic pressure model to observe the influence of those operating parameter on membrane performances. The model is evaluated by comparing with the experimental data.

2. Material and methods

Used lube oil was obtained from automotive crankcase engine in certain concentrations of impurities. The hollow-fibre polypropylene membrane module was supplied by GDP Filter, and the schematic and experimental set-up of membrane apparatus are shown in Figure 1. The samples of used oils with various concentrations of impurities were placed in a feed tank and was subjected to screen filter prior to the membrane module for minimizing module blocking from relatively bigger particles. The flux was calculated using equation 1.

$$J = \frac{Q}{A} \quad (1)$$

where, J_i is the flux of permeate (L.m^{-2}.h^{-1}), Q is volumetric rate of permeate (L), and A is the membrane surface area (m^2).

The rejection of impurities was analyzed based on the concentration of the impurities in the feed and permeate, and was calculated using equation 2.

$$R_i = \frac{C_f - C_p}{C_f} x100\% \quad (2)$$

where, R_i is the rejection of component i (%), C_{if} is concentration of component i in feed stream (wt%), and C_{ip} is the concentration of component i in permeate (%-wt.).

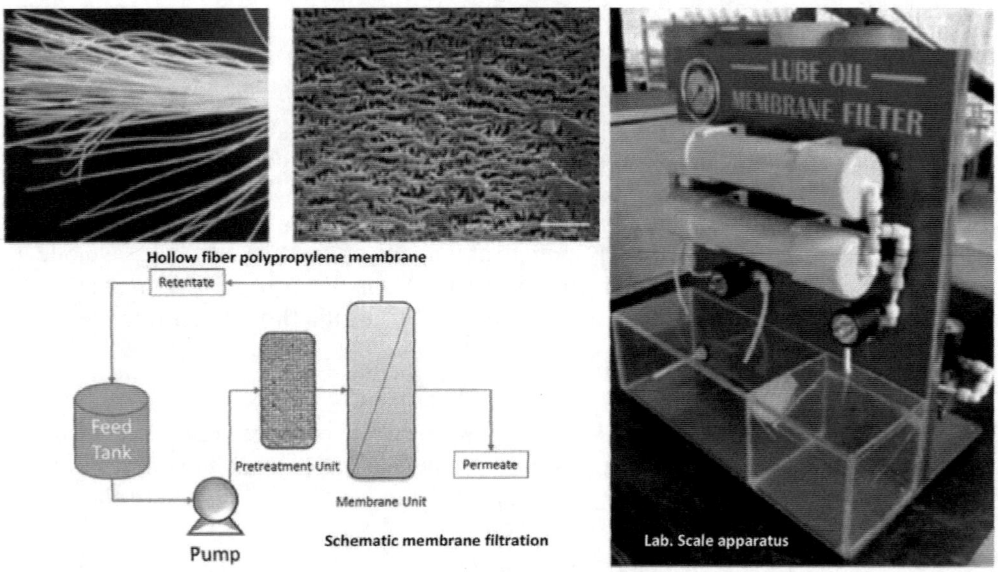

Figure 1. Experimental set-up.

3. Result and Discussion

3.1. The effect of operating parameter on membrane performance.

Generally, flux tends to increase with increasing applied pressure and decreases with increasing solute concentration. Figure 2 shows the flux and rejection as a function of applied pressure and feed concentration. The impurities concentration highly affects the viscosity of used lube oil as described in the viscosity of permeate which was lower after membrane separation due to the removal of sludge, carbonaceous particles, etc. [27]. Lai and Smith [33] reported that the permeate flux increased with the decrease of asphaltene contents in heavy oil filtration. The dependences of flux on impurities contents were related to the changes of viscosity [38].

A higher concentration of impurities leads to a higher tendency of concentration polarization and fouling. The concentration polarization and gel formation on membrane surface result in the additional mass transfer resistance of membrane filtration leading to a lower flux. In addition, this deposition of impurities onto the membrane surfaces results in a higher osmotic pressure, gel layer formation, solute adsorption, and pore blocking [36, 39].

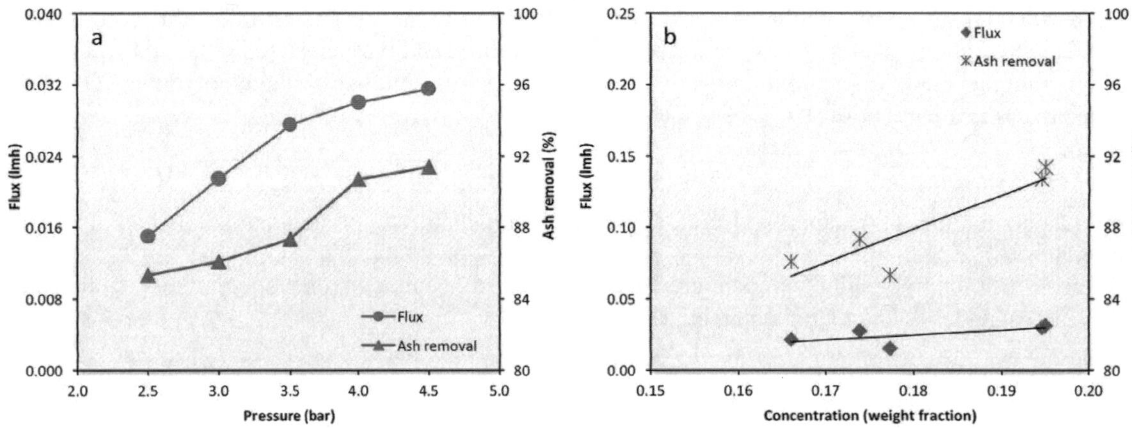

Figure 2. The Influence of applied pressure (a) and feed concentration (b) to membrane performance.

*The flux increased with the applied pressure until reach the maximum value (limiting flux), where the increasing pressure will not effectively increase the flux [36]. The value of limiting flux highly depends on the solute concentration in the feed [40]. However, according to the experimental results, the fluxes are not significantly affect*ed *by concentration in the range of* this *study. This may be due to the narrow concentration range used.*

3.2. Simulation of the membrane performance.

The influences of feed concentration and applied pressure were simulated using osmotic pressure model [40]. In the osmotic pressure model, concentration of impurities in membrane wall c_m is higher than the bulk c_b therefore the osmotic pressure of feed influences the flux of permeate, as described in equation 3.

$$J = \frac{\Delta P - \Delta\pi}{R_m} \quad (3)$$

where, ΔP is hydraulic pressure difference, $\Delta\pi$ is the osmotic pressure difference, and R_m is the membrane resistance. In a 100% rejection of solute, the concentration of solute on membrane surface c_m determines the difference of osmotic pressure. The difference of osmotic pressure shows non-linear correlation like pure solvent, as explained in equation 4.

$$\Delta\pi = a.c^n \quad (4)$$

where n constant >1, n =2 in dilute solution [41].

The flux of membrane as a function of Hydraulic pressure, osmotic pressure, feed concentration, and membrane resistance is expressed in equation 5.

$$J = \frac{\Delta P - a\, c_b^n\, exp(\frac{nJ}{k})}{R_m} \quad (5)$$

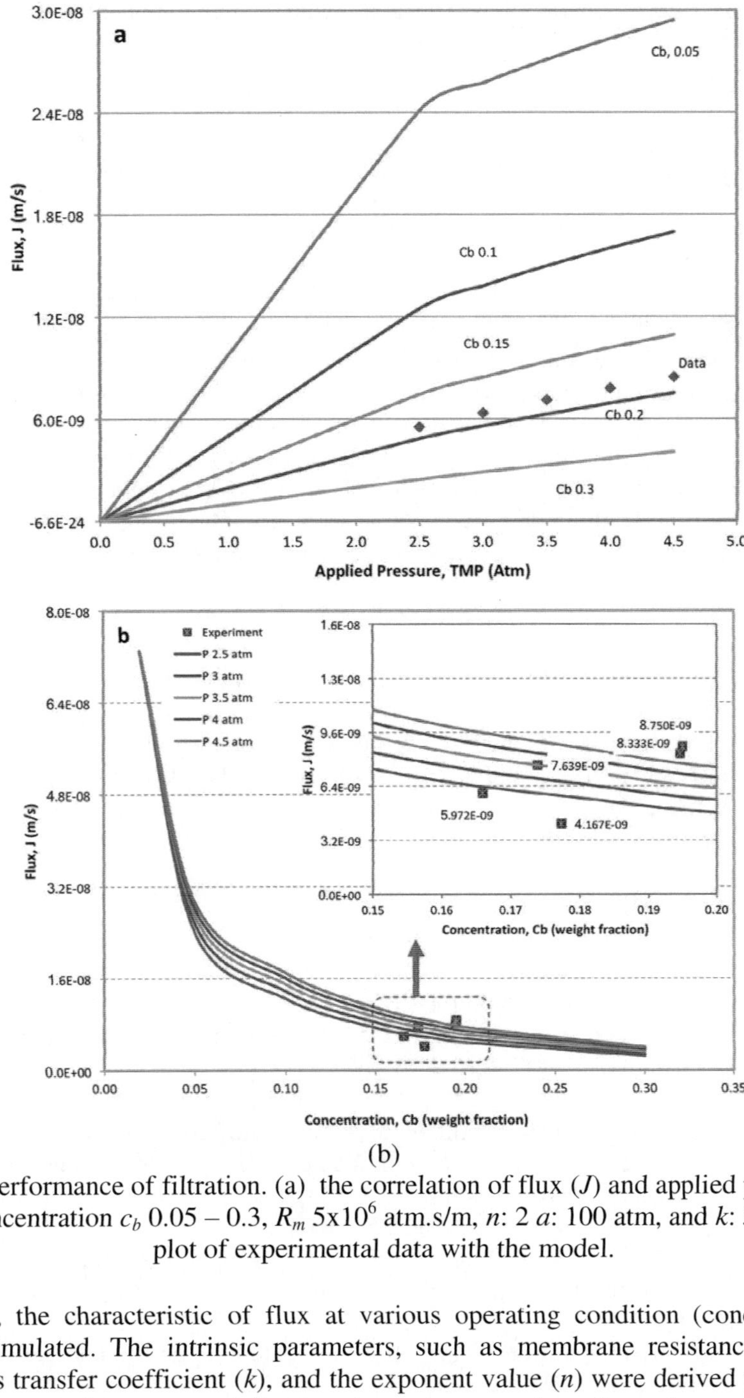

(b)

Figure 3. The performance of filtration. (a) the correlation of flux (J) and applied pressure (TMP) at various feed concentration c_b 0.05 – 0.3, R_m 5x10^6 atm.s/m, n: 2 a: 100 atm, and k: 5x10^{-9} m/s; (b) the plot of experimental data with the model.

In this work, the characteristic of flux at various operating condition (concentration, applied pressure) was simulated. The intrinsic parameters, such as membrane resistance (R_m), coefficient activity (a), mass transfer coefficient (k), and the exponent value (n) were derived using experimental data. The membrane resistance was defined as the resistance value of membrane, which was derived from measurement of steady-state flux of pure solvents through the membrane at various values of trans-membrane pressure [42]. An activity coefficient (a) is a factor used in thermodynamics to account for deviations from ideal behaviour in a mixture of chemical substances [43]. In engineering, the mass transfer coefficient is a diffusion rate constant that relates the mass transfer rate, mass transfer area, and concentration change as driving force [44]. The exponent value (n) represents the dilution of macromolecule in solution, where in semi-dilute macromolecular solutions it will have a value of about two [41], while in more concentrated solutions the exponent is even larger than two [45]. Simulation of membrane filtration was conducted using differentiation of equation 6.

$$\frac{dJ}{d\Delta P} = \left[R_m + ac_b^n\left(\frac{n}{k}\right)\exp\left(\frac{nJ}{k}\right)\right]^{-1} \tag{6}$$

The results show that the intrinsic properties of membrane filtration as stated in weight fraction value, c_b, result in the parameter of membrane resistance, $R_m = 5\text{x}10^6$ atm.s/m, the exponent value of solution, $n = 2$; $a = 100$ atm; mass transfer coefficient, $k = 5\text{x}10^{-9}$ m/s. According to those values, the simulation of membrane process to investigate the effect of concentration and applied pressure was conducted.

Figure 3 shows the influence of concentration and applied pressure to the performance of membrane filtration. The increasing of feed concentration decreases the flux of permeate, while the flux increases with applied pressure (Figure 3.a). The limiting flux is obtained at the applied pressure higher than 2.5 atm and varies with the feed concentration. The osmotic pressure builds up gradually reaching a limiting flux. At this condition, the increase in applied pressure becomes less effective or insignificant and one can define more or less arbitrarily a "limiting flux region". When the permeate flux reaches a limiting flux, the flux will be 5% of or less than the pure solvent permeability. The simulation shows a relatively good agreement with experimental data as illustrated in Figure 3.b in the range of applied pressure of 2.5 – 4.5 atm and concentration of 0.16 – 0.2 weight fraction.

4. Conclusion

In this work, the influences of solute concentration and applied pressure on the performance of hollow fiber polypropylene membrane during used lube oil filtration are investigated. Generally, the flux increases with applied pressure and decreases with concentration. By using the osmotic pressure model the intrinsic properties of the membrane, i.e. membrane resistance (R_m) and mass transfer coefficient (k) are $5\text{x}10^6$ atm.s/m and $5\text{x}10^{-9}$ m/s, respectively. The osmotic pressure model used to simulate the influence of operating parameter on membrane performances results in good agreement with experimental data.

5. References

[1] Purwasasmita M, Nabu E B P, Khoiruddin and Wenten I G 2015 Non dispersive chemical deacidification of crude palm oil in hollow fiber membrane contactor *Journal of Engineering and Technological Sciences* **47** 426-46

[2] Purwasasmita M, Juwono P B, Karlina A M, Khoiruddin K and Wenten I G 2014 Non-dissolved solids removal during palm kernel oil ultrafiltration *Reaktor* **14** 284-90

[3] Yang X, Wang R, Fane A G, Tang C Y and Wenten I G 2013 Membrane module design and dynamic shear-induced techniques to enhance liquid separation by hollow fiber modules: A review *Desalination and Water Treatment* **51** 3604-27

[4] Mynin V, Smirnova E, Katsereva O, Komyagin E, Terpugov G and Smirnov V 2004 Treatment and regeneration of used lube oils with inorganic membranes *Chemistry and technology of fuels and oils* **40** 345-50

[5] Gourgouillon D, Schrive L, Sarrade S and Rios G M 2000 An environmentally friendly process for the regeneration of used oils *Environmental science & technology* **34** 3469-73

[6] Ariono D, Khoiruddin, Subagjo and Wenten I G 2017 Heterogeneous structure and its effect on properties and electrochemical behavior of ion-exchange membrane *Materials Research Express* **4** 024006

[7] Khoiruddin and Wenten I G 2016 Investigation of electrochemical and morphological properties of mixed matrix polysulfone-silica anion exchange membrane *Journal of Engineering and Technological Sciences* **48** 1-11

[8] Xie Z, Duong T, Hoang M, Nguyen C and Bolto B 2009 Ammonia removal by sweep gas membrane distillation *Water research* **43** 1693-9

[9] Ding Z, Liu L, Li Z, Ma R and Yang Z 2006 Experimental study of ammonia removal from water by membrane distillation (MD): The comparison of three configurations *Journal of membrane Science* **286** 93-103

[10] Makertiharta I G B N, Dharmawijaya P T, Zunita M and Wenten I G 2017 Rare earth element enrichment using membrane based solvent extraction *AIP Conference Proceedings* **1805** 070001

[11] Khoiruddin, Aryanti P T P, Hakim A N and Wenten I G 2017 The role of ion-exchange membrane in energy conversion *AIP Conference Proceedings* **1840** 090006

[12] Himma N F and Wenten I G 2017 Surface engineering of polymer membrane for air separation *AIP Conference Proceedings* **1840** 090005

[13] Yeon S-H, Lee K-S, Sea B, Park Y-I and Lee K-H 2005 Application of pilot-scale membrane contactor hybrid system for removal of carbon dioxide from flue gas *Journal of Membrane Science* **257** 156-60

[14] Aryanti P, Joscarita S R, Wardani A K, Subagjo S, Ariono D and Wenten I G 2016 The influence of PEG400 and acetone on polysulfone membrane morphology and fouling behaviour *Journal of Engineering and Technological Sciences* **48** 135-49

[15] Aryanti P T P, Yustiana R, Purnama R E D and Wenten I G 2015 Performance and characterization of PEG400 modified PVC ultrafiltration membrane *Membrane Water Treatment* **6** 379-92

[16] Camacho L M, Dumée L, Zhang J, Li J-d, Duke M, Gomez J and Gray S 2013 Advances in membrane distillation for water desalination and purification applications *Water* **5** 94-196

[17] Arar Ö, Yüksel Ü, Kabay N and Yüksel M 2013 Application of electrodeionization (EDI) for removal of boron and silica from reverse osmosis (RO) permeate of geothermal water *Desalination* **310** 25-33

[18] Wood J, Gifford J, Arba J and Shaw M 2010 Production of ultrapure water by continuous electrodeionization *Desalination* **250** 973-6

[19] Khoiruddin K, Ariono D, Subagjo and Wenten I G 2017 Surface modification of ion-exchange membranes: Methods, characteristics, and performance *Journal of Applied Polymer Science* **134** 45540

[20] Wenten I and Khoiruddin 2016 Recent developments in heterogeneous ion-exchange membrane: preparation, modification, characterization and performance evaluation *Journal of Engineering Science and Technology* **11** 916-34

[21] Wardani A K, Hakim A N, Khoiruddin and Wenten I G 2017 Combined ultrafiltration-electrodeionization technique for production of high purity water *Water Science and Technology* **75** 2891-9

[22] Aryanti P, Sianipar M, Zunita M and Wenten I 2017 Modified membrane with antibacterial properties *Membrane Water Treatment* **8** 463-81

[23] Ariono D, Purwasasmita M and Wenten I G 2016 Brine effluents: Characteristics, environmental impacts, and their handling *Journal of Engineering and Technological Sciences* **48** 367-87

[24] Adham S, Hussain A, Matar J M, Dores R and Janson A 2013 Application of membrane distillation for desalting brines from thermal desalination plants *Desalination* **314** 101-8

[25] Wenten I G, Julian H and Panjaitan N T 2012 Ozonation through ceramic membrane contactor for iodide oxidation during iodine recovery from brine water *Desalination* **306** 29-34

[26] Dermentzis K 2010 Removal of nickel from electroplating rinse waters using electrostatic shielding electrodialysis/electrodeionization *Journal of Hazardous Materials* **173** 647-52

[27] Cao Y, Yan F, Li J, Liang X and He B 2009 Used lubricating oil recycling using a membrane filtration: Analysis of efficiency, structural and composing *Desalination and Water Treatment* **11** 73-80

[28] Merai Yash P 2015 Re-refining of used lubricating oil *International Journal of Scientific & Engineering Research* **6** 329-32

[29] Srivastava S 2014 Rerefining and Recycling of Used Lubricating Oil - *Developments in Lubricant Technology* (John Wiley & Sons, Inc.) pp 299-307

IOP Conf. Series: Materials Science and Engineering **395** (2018) 012018 doi:10.1088/1757-899X/395/1/012018

[30] White L S and Nitsch A R 2000 Solvent recovery from lube oil filtrates with a polyimide membrane *Journal of Membrane Science* **179** 267-74
[31] Wenten I G 2002 Recent development in membrane science and its industrial applications *J Sci Technol Membrane Sci Technol* **24** 1010-24
[32] Dahlan M H, Setiawan A and Rosyada A 2014 Pemisahan oli bekas dengan menggunakan kolom filtrasi dan membran keramik berbahan baku zeolit dan lempung *Jurnal Teknik Kimia* **20** 38-45
[33] Lai W-C and Smith K 2001 Heavy oil microfiltration using ceramic monolith membranes *Fuel* **80** 1121-30
[34] Psoch C, Wendler B, Goers B, Wozny G and Ruschel B 2004 Waste oil conditioning via microfiltration with ceramic membranes in cross flow *Journal of membrane science* **245** 113-21
[35] Siswanti 2010 Pengaruh penambahan aditif proses daur ulang minyak pelumas bekas terhadap sifat-sifat fisis *Eksergi* **10** 27-31
[36] Baker R W 2012 *Membrane Technology and Applications* (John Wiley & Sons, Inc.)
[37] Mohammadi T and Safavi M A 2009 Application of Taguchi method in optimization of desalination by vacuum membrane distillation *Desalination* **249** 83-9
[38] Duong A, Chattopadhyaya G, Kwok W Y and Smith K J 1997 An experimental study of heavy oil ultrafiltration using ceramic membranes *Fuel* **76** 821-8
[39] Mulder J 2012 *Basic Principles of Membrane Technology* (Springer Science & Business Media)
[40] Wijmans J, Nakao S and Smolders C 1984 Flux limitation in ultrafiltration: osmotic pressure model and gel layer model *Journal of membrane science* **20** 115-24
[41] De Gennes P-G 1979 *Scaling Concepts in Polymer Physics* (Cornell university press)
[42] Chisti Y 2007 Principles of membrane separation processes *Bioseparation and bioprocessing: a handbook, 2nd ed.* (Wiley-VCH, New York)
[43] McNaught A D and McNaught A D 1997 *Compendium of Chemical Terminology* vol 1669 (Blackwell Science Oxford)
[44] Seader J D, Henley E J and Roper D K 1998 *Separation Process Principles* (John Wiley & Sons, Inc.)
[45] Flory P J 1953 *Principles of Polymer Chemistry, 2nd ed.* (Cornell University Press)

IOP Conf. Series: Materials Science and Engineering **395** (2018) 012019 doi:10.1088/1757-899X/395/1/012019

Photodegradation Effect of Structure of Linear Low Density Polyethylene – Starch – Clay Nanocomposite Film

C Liza[1*], B Soegijono[2], E Budianto[2], Syuhada[1] and D Rusmana[1]

[1] Balai Teknologi Polimer, Puspiptek Office Area, Bld. 461, Tangsel-Banten, Indonesia
[2] Department of Physics, Material Science, University of Indonesia, Depok - Indonesia

*Email: otisica@yahoo.com

Abstract. Degradable plastic is a topic that continues to be discussed, especially for disposable applications. Blending starch with commercial plastics such as Linear Low Density Polyethylene was the alternative that continues to be developed. The degradation mechanism of biopolymer may occur as microorganisms and also through the mechanism of exposure to sunlight or Ultra Violet light. In this work, degradation test on Linier Low Density Polyethylene-starch-clay nanocomposite film was conducted with a Xenon Arc. Nanocomposite samples were prepared from masterbatch clay and 10% wt and 20% wt of starch. Pure Linear Low Density Polyethylene was used as a control parameter. On day 7th, 20% wt starch sample was already brittle and 10%wt starch was brittle on day 14th. In the 3th day of Xenon Arc aging, starch granulates already disappeared on the surface of sample and the degraded starch leave a hole on the surface of Linear Low Density Polyethylene matrix. The hole dimension became larger with time increasing of Xenon Arc aging. Thermo Gravimetric Analysis result showed decreasing onset temperature degradation for the starch. The oxidation and degradation of starch was occurred during the aging. Meanwhile onset temperature degradation of Linear Low Density Polyethylene in the nanocomposite sample was increased might be this occurred due to the formation of crosslink on Linear Low Density Polyethylene before degraded. Image of 10% wt and 20% wt starch observed using Atomic Force Microscope also showed that the degradation occurs on starch. It was proved by the image after 14 days aging become the same as the image of Linear Low Density Polyethylene used as a control that showed the lamella of Linier Low Density Polyethylene.

1. Introduction

The disposable plastic packaging was the largest contributor to waste production. To reduce the problem of plastic waste, a lot of research done on plastic degradable either use of natural resources or the use of additives such as pro-oxidant [1-5]. Starch was one of the natural resource that was developed in packaging material. However, the addition of starch could reduce mechanical, barrier properties and transparency [1, 3]. Reinforce nanoclay was used to enhance mechanical and barrier properties of starch plastic.

Blending starch with commercial plastic will improve degradability [1,4]. Degradation in polymer material can be stimulated by different factor, such as light, moisture, heat and microorganism. Photodegradation generates material to undergo degradation mechanism with atmospheric oxygen in the presence of light [6-8]. Photodegradation can be done with direct sunlight exposure method

IOP Conf. Series: Materials Science and Engineering **395** (2018) 012019 doi:10.1088/1757-899X/395/1/012019

(outdoor) or using accelerated weathering chamber (indoor), used to exposed material at accelerated atmospheric condition [6,8]. For outdoor exposure test, needed long time period to complete the experiment, compare to indoor method that faster, easier and expected close to real condition or similar to outdoor exposure.

For further analysis of photodegradation is necessary to evaluate the result. Several analysis were performed after photodegradation such as, color changes, mechanical, chemical and thermal properties [6,8]. The objective of this research was to characterized morphological structure of degradable LLDPE starch clay nanocomposites after photodegradation.

2. Experimental

2.1. Material

Clay used was organo layer silicate (OLS) cloisite 20A from Southern Clay that had been treated with glycerol monostearate (GMS). Glycerol monostearate was provided by anonymous local company. Ethanol was used as a solvent to dissolve GMS. Polyethylene graft maleic anhydride (PEgMA) from Sigma Aldrich and cassava starch from commercial market was used as a compatibilizer and biodegradable material. LLDPE Asrene UF 1810, PT. Chandra Asri for blown film application was used as polymer matrix

2.2. Methods and Characterization

PEgMA, OLS treated with GMS and cassava starch was masterbatching with Rheomix Haake refer to our previous work [9]. Masterbatch and LLDPE were dry mixed before feeding to twin screw extruder, ZK 25 E-Collin to produce compound at 40 – 170°C and 20 rpm. Then, the compound was dried in an oven vacuum for 24 hours at 50°C. Sample film was prepared by Blown Film Rheomix Haake. Degradation process of sample film was conducted using Xenon Arc Ci 3000 at the condition: irradiant 0.55 ± 0.01 W/m2 @ 340 nm, a single cycle run for 4.8 hours with 1 hour dark and 3.8 lights. The condition of 1 hour dark were at temperatures of 38°C; RH: 95% and 3.8 hours light were at temperatures of 89°C and RH of 50% for 14 days (fig.1). Thermal gravimetric analysis (TG 209 Tarsus, Netzch) was performed at 50 – 600°C under nitrogen gas atmosphere.

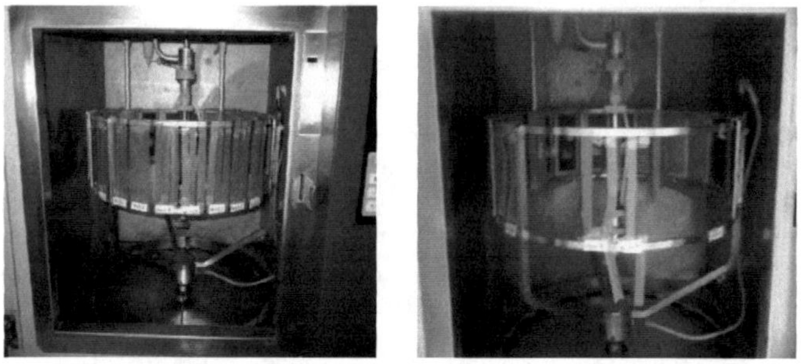

(a) (b)

Figure 1. Photodegradation using Xenon Arc: (a) Before aging and (b) After 14 days aging.

Samples surface after treated in Xenon Arc were characterized with SEM (JSM-6510LA, Jeol) with a magnification of 500 times. The samples were coated with thin layer of platinum. The surface morphological of thin layer sample was also examined using Atomic Force Microscope (AFM-5500, Agilent). Samples were placed between two silicon wafers, heated at 120°C and then pressed to form a thin layer prior to AFM preparation test. The AFM observation by mean topographic and phase images analysis were taken simultaneously, perform in non-contact mode and scan size 5x5μ. Fourier Transform Infra Red (FTIR) spectroscopy (Bruker, Tensor 27) were conducted to observe the degradation that occurs. Samples measured with free standing film method.

3. Result and Discussion

Figure 1.a showed film samples before aging with xenon Arc and figure 1.b showed most film samples were already degraded. Observation result after photodegradation in elongation at break of 10% wt and 20%wt Starch-LLDPE-Clay Nanocomposite decreased significantly [10].

Figure 2 and 3 showed morphological surface of 10% wt and 20% wt Starch-LLDPE-Clay Nanocomposite before and after Xenon Arc photodegradation, characterized with SEM. Its showed that before aging, the starch granulates were observed at the sample surface. In the 3th day of Xenon Arc aging, starch granulates already disappeared on the surface of sample and the degraded starch leave a hole on the surface of LLDPE matrix. The hole dimension became slightly larger with time increasing of Xenon Arc aging. In 20%wt starch, the hole dimension of degraded starch had larger than the 10% wt starch. LLDPE control sample without starch and clay showed no changes after photodegradation with Xenon Arc (fig.4).

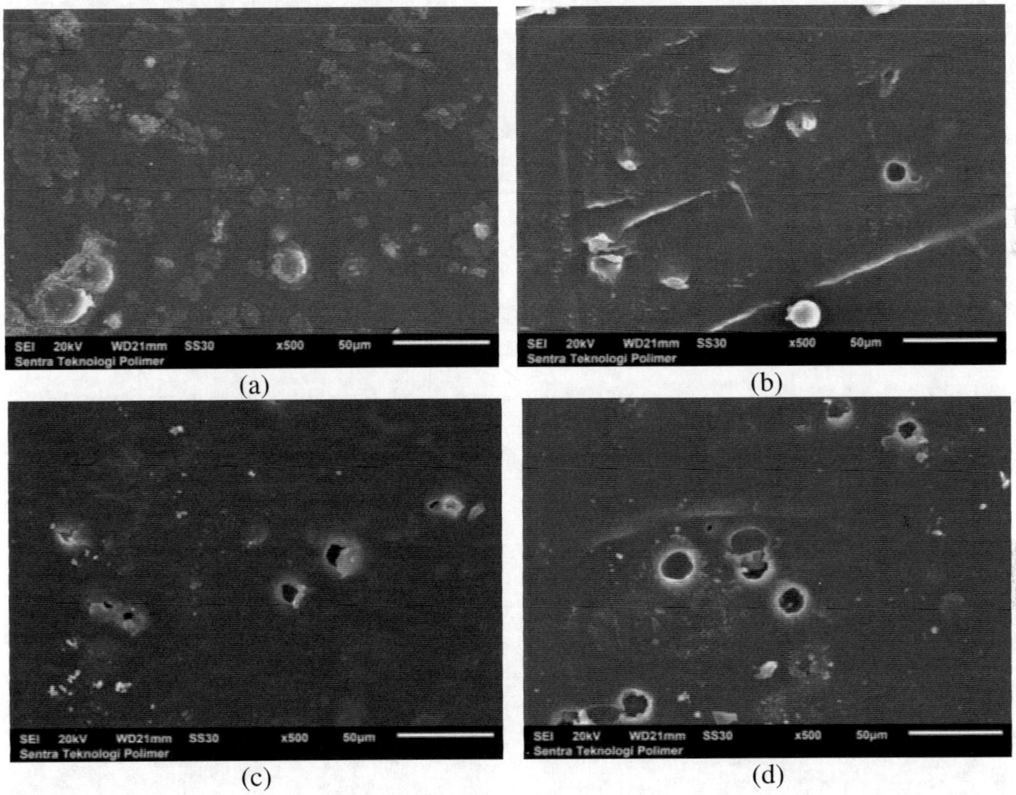

Figure 2. Morphological structure of 10%wt Starch-LLDPE-Clay Nanocomposite by SEM, before and after Xenon Arc photodegradation:
(a) before, (b) after 3 days, (c) after 7 days and (d) after 14 days.

IOP Conf. Series: Materials Science and Engineering **395** (2018) 012019 doi:10.1088/1757-899X/395/1/012019

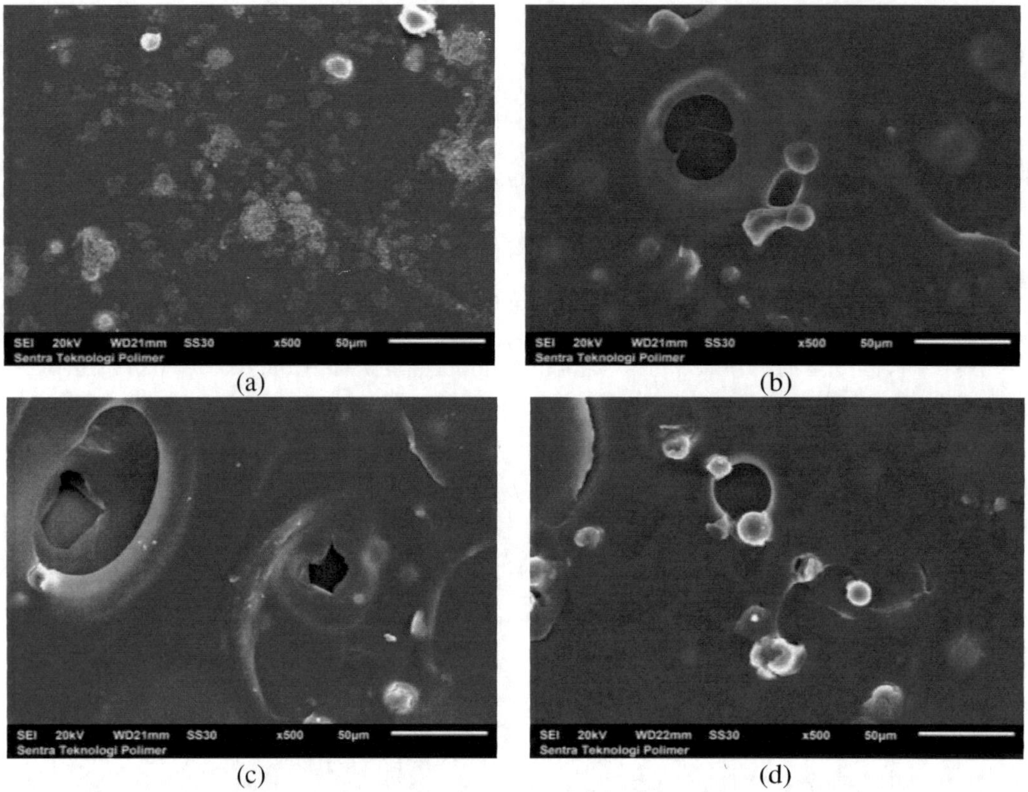

Figure 3. Morphological structure of 20%wt Starch-LLDPE-Clay Nanocomposite by SEM, before and after Xenon Arc photodegradation:
(a) before aging, (b) after 3 days, (c) after 7 days and (d) after 14 days.

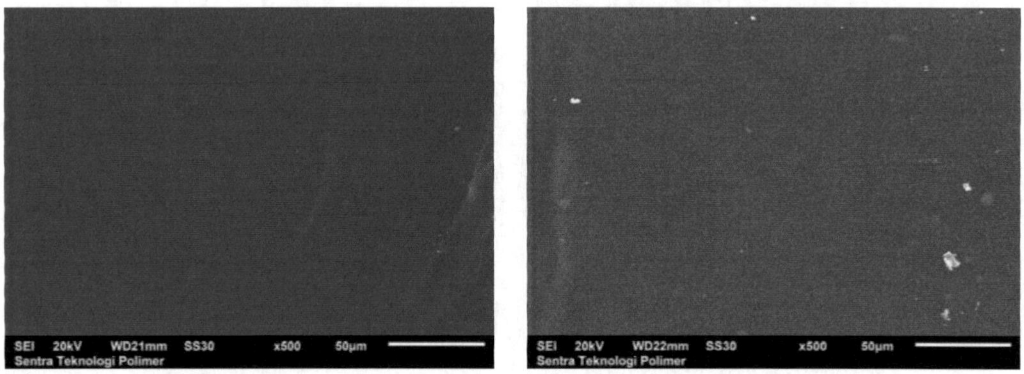

(A) (B)

Figure 4. Morphological structure of LLDPE control by SEM, before and after Xenon Arc photodegradation: (a) before and (b)after 14 days.

AFM images showed fine spherical particle on surface structure of 10% wt and 20% wt Starch-LLDPE-Clay Nanocomposite before aging. This particle was blocklet of starch, similar to the result of reported by Park et al., 2009 that was 50-200 nm blocklets of starch [11,12]. Fig. 5.b1 showed the blocklet dimension of 10%wt starch about 200 nm meanwhile the blocklet dimension of 20% wt starch was about 100 nm. LLDPE before aging showed lamella structure with thickness of lamella was about 60 – 80 nm (fig.5.a1). Topograpy of lamella structure characterized with AFM was refer to

IOP Conf. Series: Materials Science and Engineering **395** (2018) 012019 doi:10.1088/1757-899X/395/1/012019

Chen et al., 2011 and Golebiewski et al. 2008 [13,14]. There was no significant changes in lamella structure of LLDPE before and after 14 day aging with Xenon Arc. Morphological structure surface of 10% wt and 20% wt Starch-LLDPE-Clay Nanocomposite after 14 days aging showed the similar structure of LLDPE control, but the diameter of lamella more smaller (table 1). This shows that the starch has been degraded by no longer visible on the surface of the sample as well as changes in LLDPE lamella size. The degradation, further observed by FTIR spectroscopy measurements.

Table 1. Morphology of 10% wt, 20% wt Starch-LLDPE-Clay Nanocomposite film and LLDPE control, before and after 14 days Xenon Arc.

Sample	Aging	Blocklet/lamella diameter (nm)	Possible structure
10 % wt starch	Before Aging	± 200	Starch blocklet at surface
	Aging 14 days	40 - 80	Lamella LLDPE
20 % wt starch	Before Aging	± 100	Starch blocklet at surface
	Aging 14 days	40 - 60	Lamella LLDPE
LLDPE control	Before Aging	60 - 80	Lamella LLDPE
	Aging 14 days	60 - 80	Lamella LLDPE

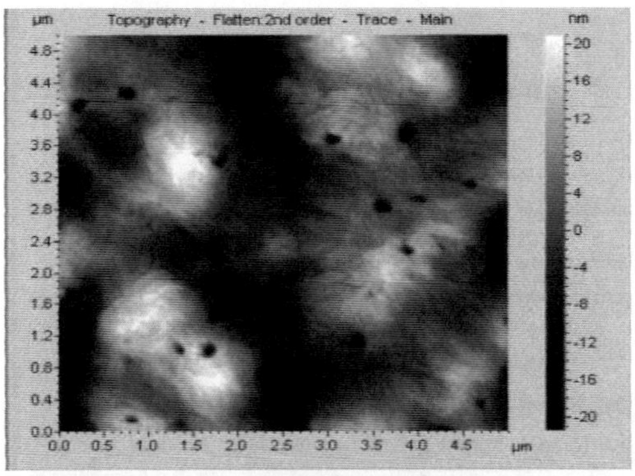

(a.1)

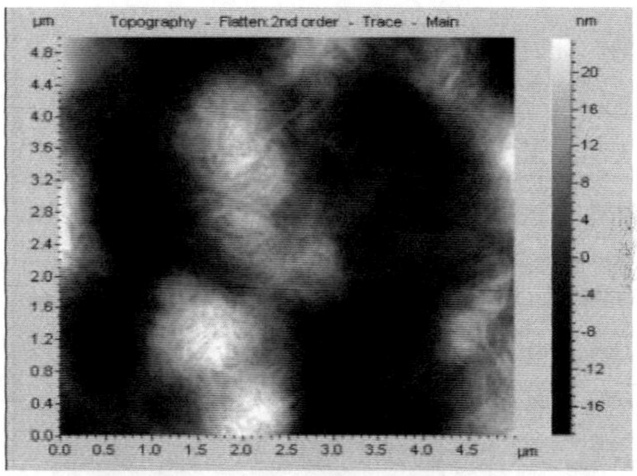

(a.2)

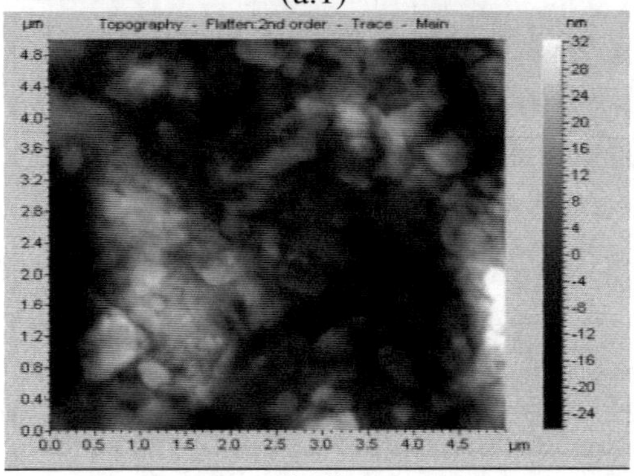

(b.1)

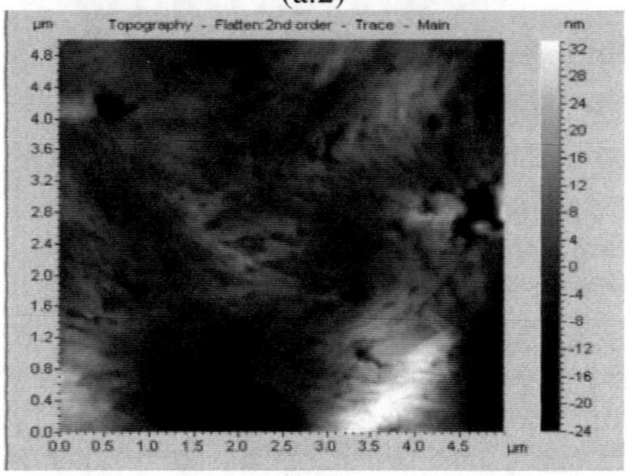

(b.2)

IOP Conf. Series: Materials Science and Engineering **395** (2018) 012019 doi:10.1088/1757-899X/395/1/012019

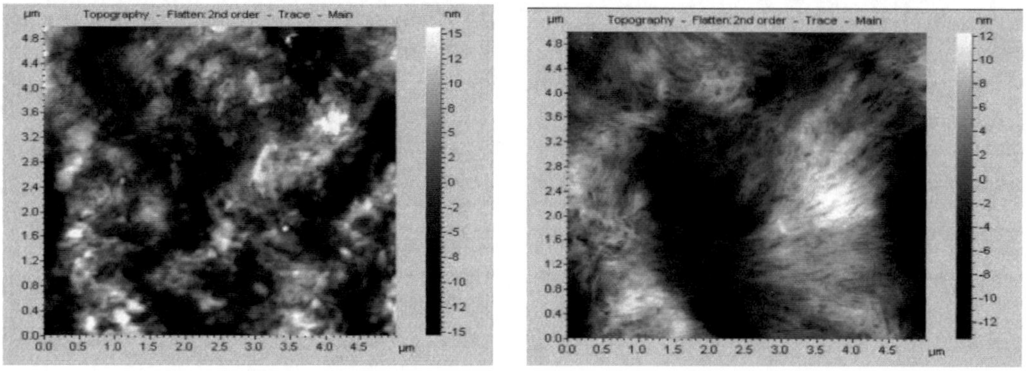

(c.1) (c.2)

Figure 5. Morphological structure by AFM (tophography), before and after Xenon Arc photodegradation: (a) LLDPE control, (b) 10%wt Starch -LLDPE-Clay Nanocomposite and (c) 20% wt Starch-LLDPE-Clay Nanocomposite; (1) before aging and (2) after 14 days.

Figure 6 showed FTIR spectroscopy result for nanocomposite 10% wt and 20% wt starch proved that degradation was occurred by increasing the intensity in the wave number at 1722 cm^{-1}, corresponding to carbonyl bond [4]. The longer the degradation, the intensity of carbonyl bond was higher. The oxidation rate of LLDPE due to photodegradation was proportional to amorphous region and the degree of crystallinity will increase. Crystallinity bands arising can be observed (fig. 6) in the splitting of the CH_2 rocking band into a pair of bands at 732 cm^{-1} and 719 cm^{-1}.

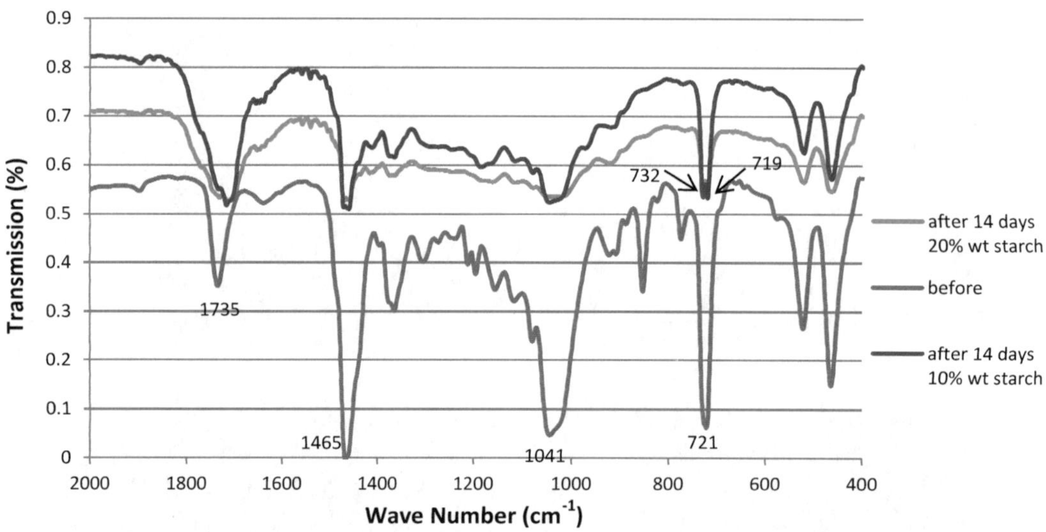

Figure 6. FTIR spectrum of 10 and 20 %wt starch, before and after 14 days Xenon Arc.

Thermo Gravimetric Analysis result (table 1) showed decreasing onset degradation temperature for the starch in nanocomposite. The oxidation and degradation of starch was occurred during the aging. Meanwhile onset temperature of LLDPE in nanocomposite sample was increased, this occurred due to the high energy radiation that lead the formation of chain scission or crosslink on LLDPE before degraded. Image of 10%wt and 20%wt starch observed using Atomic Force Microscope also showed that the degradation occurs on starch. It was proved by the image after 14 days aging become the same as the image of LLDPE used as a control that showed the lamella of LLDPE.

Table 2. Onset Degradation Temperature of 10% wt and 20% wt Starch in Nanocomposite film, before and after 14 days Xenon Arc.

Sample		Onset Degradation Temperature (^{o}C)	
		Before Aging	Aging 14 days
10 % wt starch	Starch	282.6	266.6
	LLDPE	447.2	450.2
20 % wt starch	Starch	279.3	259.0
	LLDPE	440.9	445.4

4. Conclusion

Degradation at 10%wt and 20%wt Starch in nanocomposite film sample were occur initially at starch. There were no changes in lamella of control LLDPE without starch, after 14 days photodegradation with Xenon Arc. SEM analysis showed a hole in surface nanocomposite film sample causes by starch granulate degradation. FTIR and TGA result also proved that degradation occurred at starch and LLDPE in nanocomposite film sample.

5. References

[1] Ruxanda B, Iuliana S and Carmen Alice T 2010 Enzymatic Degradation of LDPE / Corn Starch Treated with [EMIM][Cl] Ionic Liquid *Materiale Plastice* **47** 126
[2] Hongseng L, Fengwei X, Long Y, Ling C and Lin L 2009 Thermal Processing of Starch-Based Polymers *Progress in Polymer Science* **34** 1348
[3] D R Lu, Congming X and Shanjun X 2009 Starch-Based Completely Biodegradable Polymer Materials *eXPRESS Polymer Letters* **3** 366
[4] Leon P B M J and Leszek M 2006 Thermoplastic Starch as Packaging Material *Acta Sci. Pol. Technica Agararia* **5 (1)** 19
[5] Maurizio A, Jan De V, Maria E E, Sabine F, Paulo V and Maria G V 2005 Biodegradable Starch/Clay Nanocomposite Films for Food Packaging Applications *Food Chemistry* **93** 467
[6] Anne A et.al. 2011 An overview of degradable and biodegradable polyolefins *Progress in Polymer Science* **36** 1015
[7] Pham N L, Nguyen C G and Nguyen H T 2007 Photodegradable Films Based on Polyethylene-Cassava Starch Blend Using Benzophenone/Antraquinone or Fe(II) Compound as Photosensitive Agents *Journal of Chemistry* **45** 392
[8] Lucas N, Bienaime C, Belloy C, Queneudec M, Silvestre F and Nava-Saucedo J E 2008 Polymer biodegradation: Mechanisms and estimation techniques *Chemosphere* **73** 429
[9] Chandra L, Bambang S, Emil B, Syah Johan A dan Jayatin 2013 Morphological Structure of Pretreated Organo Layer Silicate Dispersion in Masterbatch Biodegradable Polymer Cassava Starch Clay Nanocomposite *Adv. Materials Research* **626** 157
[10] Chandra L, Bambang S, Emil B, Jayatin and Saeful R 2013 Accelerated Degradation of LLDPE-Starch-Clay Nanocomposite Film with Xenon Arc and Characterized on Mechanical and Chemical Properties *15th Asian Chemical Congress*
[11] So H P, K.C. Eom and Sung Y Y 2009 Nano-structural Characteristics of Gelatinized Starch Granule in Pigmented Rice Seeds *The 9th Esafs*
[12] U.V. Lay Ma, John D F and Gregory R Z 2011 Effect of Starch Fractions on Spherulite Formation and Microstructure *Carbohydrates Polymer* **83** 1757
[13] Jan G, Artur R, Janus D and Andrzej G 2008 Low Density Polyethylene–Montmorillonite Nanocomposites for Film Blowing *European Poly. J* **44** 270

IOP Conf. Series: Materials Science and Engineering **395** (2018) 012019 doi:10.1088/1757-899X/395/1/012019

[14] Ke W, Feng C, Qin Z and Qiang F 2008 Shish–Kebab of Polyolefin by ''Melt Manipulation'' Strategy in Injection-Molding: A Convenience Pathway from Fundament to Application", *Polymer* **49** 4745

Acknowledgement
Financial support from Balai Teknologi Polimer and Ministry of Research Science, Technology and Higher Education of Indonesia are gratefully acknowledged. The author would like to thank to Institute Technology of Indonesia for providing mixing with Rheomix Haake.

IOP Conf. Series: Materials Science and Engineering **395** (2018) 012020 doi:10.1088/1757-899X/395/1/012020

Evaluation of the Effect of *Curcuma longa L.* Essential Oil in Chitosan-starch Edible Coating

N M Yusof[1,*], J Jai[1], F Hamzah[1] and S Pinijsuwan[2]

[1]Faculty of Chemical Engineering, Universiti Teknologi MARA, 40450 Shah Alam, Selangor, Malaysia.
[2]School of Agro Industry, Mae Fah Luang University, 57100 Chiang Rai, Thailand.

*noorsuhana@salam.uitm.edu.my

Abstract. *Curcuma longa L.* is one of the commercially planted herbs in Malaysia for its phytochemical benefits. In this study, the effect of chitosan-starch at different concentration of *Curcuma longa L.* (CUR) were examined. The application of edible coating by using dipping technique has been analysed for their effectiveness in extending shelf life of fruits through several analyses include the microstructure observation, disk diffusion and appearance. Chitosan-starch coating incorporated with 10 uL of CUR had the best appearance and resulted a layer of homogeneous surface about 37.45 μm as can be seen in microstructure observation. The disk diffusion analysis proves that the chitosan with 10uL of curcuma longa had the highest inhibition zone which is 17 mm. Based on the present study, chitosan-starch with the addition of (CUR) is believed to have a major potential in extending shelf life of fruits with high antimicrobial activity from local herbs.

1. Introduction

Strawberries are very sensitive to high temperature that can be easily damage due to the handling processes once they were harvested. It can lead to produce rotten fruits at very minimal time even though the right storage utilized on them. Several methods have been proposed to overcome this including the application of technology by physical, chemical and biological means to prevent deterioration. By applying physical treatment like drying for example, lots of drawback may arise from this treatment such as the heat being applied may alter the quality of product such as nutrient loss, change in texture, colour and odour. Other than that, if chemical treatment being applied the main concern among all these synthetic chemicals is the residues that will remains in the product and can be very harmful. Thus, the application of edible coating enriched with essential oils using curcuma longa can be used as alternative to prolonging shelf life of fruits. Application of edible film from chitosan and starch enriched with curcuma longa essential oil can hold off the actual deterioration of strawberries as well as keep up with the freshness for a longer period.

Chitosan is derivation of chitin component via deacytylation in alkaline media and it is a natural, nontoxic copolymer consist of β- (1-4)-2-acetamido-D-glucose and β-(1-4)-2-amino-D-glucose units. It is abundant as it naturally can be found in the exoskeleton of crustaceans, fungal cell walls and other biological materials [1]. Chitosan coating preserve the quality of fruit by regulating gas exchange and decrease the respiration. It reduces the oxidative process due to reduction of reactive oxygen species accumulation by lowering the respiration rate [2]. It has been proven by applying the chitosan as coating on postharvest fruit including logan, pear, plum and apple to maintain its quality [3].

Edible coating is a thin layer of materials that offers a barrier to oxygen, microbes that come from environment, moisture loss and solute movement for food plus, it can be consumed by consumers [4]. It also does not add unfavourable properties to the fruits and environmental friendly due to its natural biodegradable source. It works by delaying the rate of respiration, decrease weight loss and extend the shelf life of fruits and vegetable during the postharvest storage [5]. The addition of essential oil would give significant effect on antimicrobial activity against bacteria, yeast and moulds and higher safety to environment and consumer [6].

The use of *Curcuma longa L.* (CUR) essential oil which abundantly available in Malaysia as one of the natural antimicrobial meet the needs. According to Wang [7], the antimicrobial, antibacterial and antifungal activities of CUR against foodborne pathogens are very promising to be use as the preservative in food industry. These properties attributed from the existence of phenolic compounds composed mainly of curcumin, demethoxy curcumin and bis-demethoxy curcumin [8].

2. Materials and method

2.1. Materials

Starch was purchased from local market, *Curcuma longa L.* being used as the essential oil was bought from BF1 Sdn. Bhd., 0.1M sodium hydroxide, Tween 80, acetic acid glacial and glycerol were bought from R&M Chemicals while medium molecular weight chitosan was purchased from Sigma Aldrich.

2.2. Edible coating preparation

Chitosan-starch edible coating was developed based on the procedures by [9] with some modifications.

2.2.1. Chitosan solution. Chitosan with 1% (w/v) was gelatinized in an aqueous solution of 0.5% (w/v) acetic acid at room temperature. 0.1 % (w/v) of Tween 80 was added as a surfactant to increase wettability [10]. To ensure that the chitosan is completely dissolved, the solution was stirred for 24 hours at 350 rpm until all the chitosan flake disappear. pH of the solution was adjusted by using 0.1 M sodium hydroxide until it reach pH of 5.6 [1] to decrease the acidic condition of chitosan solution. The solution was made up to 100 mL by adding distilled water up to 100 mL.

2.2.2. Starch solution. Starch solution at 4% (w/v) was gelatinized at 80°C. The solution was stirred for 30 minutes at 350 rpm to overcome the starch from being coagulated. The solution was made up to 100 mL by add up distilled water until reach to 100 mL. To enhance the application of solution formed by the combining chitosan and CUR essential oil, 2 mL of glycerol was added as a plasticizer.

2.2.3. Chitosan and starch coating solutions. Prepared starch solution at 2% (w/v) was added to 100 mL of prepared chitosan solution. The combined solution was stirred at room temperature for 18h to enable the solution well mixed. For the combination of CUR into chitosan and starch, different concentration of CUR essential oil was used which are 20, 10, 5, and 0 µL. The CUR essential oil was added into the prepared chitosan and stirred at room temperature until complete mixing. The coating solution formulations were added with chitosan concentrations of: 20, 15, 10 and 5 µL; glycerol concentrations of 2.0% and (v/v); Tween 80 concentrations of 0.1% (w/v).

2.2.4. Chitosan-starch with CUR essential oil solution. 100 mL of chitosan solution was taken, and 2 mL of prepared starch solution was added into the chitosan solution. After that, 5 µ L of CUR was

IOP Conf. Series: Materials Science and Engineering **395** (2018) 012020 doi:10.1088/1757-899X/395/1/012020

added to the mixture. The process was repeated for different concentration of CUR which is 10, 15 and 20 μ L for every 100 mL of prepared chitosan and 2 mL of starch solution [11].

2.3. Microstructure observation

The microstructure of coated sample was analysed using a field emission-scanning electron microscopy (FESEM) Zeiss Supra 35VP. The strawberry sample were mounted onto metal stubs and coated with Au using an accelerating voltage of 5 kV.

2.4 Disk diffusion analysis

A sterile swab was used to take fungal culture from strawberry sample. Then, the culture was streak onto the Potato Dextrose Agar plate. The plate was steak with the swab in one direction and then rotate the 90° and streak the plate again in that direction to obtain uniform growth. The rotation was repeated for 3 times. After that, antibiotic disc that had been immersed with the chitosan samples of different concentration of turmeric oil were dispensed onto the agar. The forceps were used to gently press each of the disc to the agar to ensure each of the disc was attached to the agar. The forceps were sterilized by using flame. The plates were incubated at 25°C for 48-72 hours [12].

2.5 Physical appearance

The laboratory assays were done using fresh strawberries with same the maturity stage, 75% red colour, uniform weight with no sign of mechanical or microbiological injury.

3. Results and discussion

3.1. Microstructure observation

Field emission scanning electron microscopy (FESEM) was carried out to observe the microstructure and surface characteristics of coating film to establish a relationship between coating and strawberry. Figure 1 shows FESEM micrograph of cross section of strawberry coated with chitosan-starch coating plus 10 μL of CUR essential oil. As can be seen the coating had a quite smooth and homogeneous surface with a thickness about 37.45 μm. It suggests that the blending of chitosan and starch with the addition of CUR essential oil are highly miscible and homogeneous.

3.2. Disk diffusion analysis

The inhibition zone was recorded through disk diffusion analysis as can be seen in Figure 2. The inhibition zone indicates the antifungal properties for each of the chitosan coating that had been added with different concentration of CUR essential oil. After three day of incubation period, the inhibition zone for chitosan coating added with 10 ul of turmeric oil showed the highest inhibition zone which is 17 mm while the chitosan coating with 5ul turmeric oil give 10mm. For other samples, there were no sign of antifungal activity. Chitosan-starch coating added with 10ul of turmeric oil has the highest antifungal properties thus prolong the shelf life of strawberry samples. As proven by the investigation by Perdones [13], the use of chitosan coating would result to reduction in growth of Botrytis cinerea thus, it showed the degree of antimicrobial activity. Moreover, curcumin found in CUR has fairly good antimicrobial effect against large number of microorganisms and give a good antiseptic effect when the concentration is above 0.05% [14], [15].

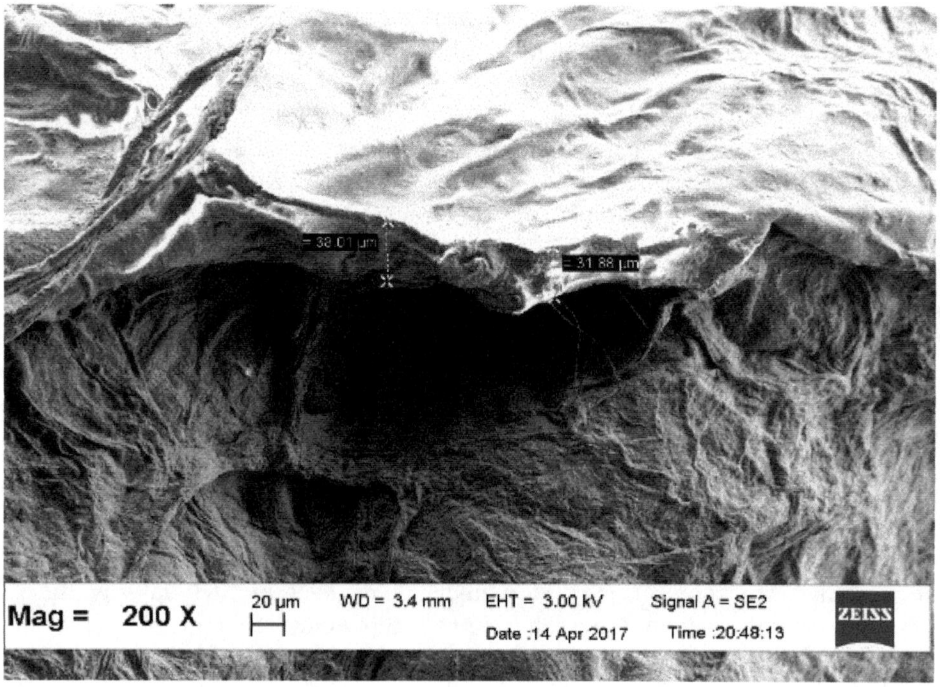

Figure 1. FESEM micrographs of the cross section of the strawberry coated with chitosan-starch based coating containing CUR essential oil.

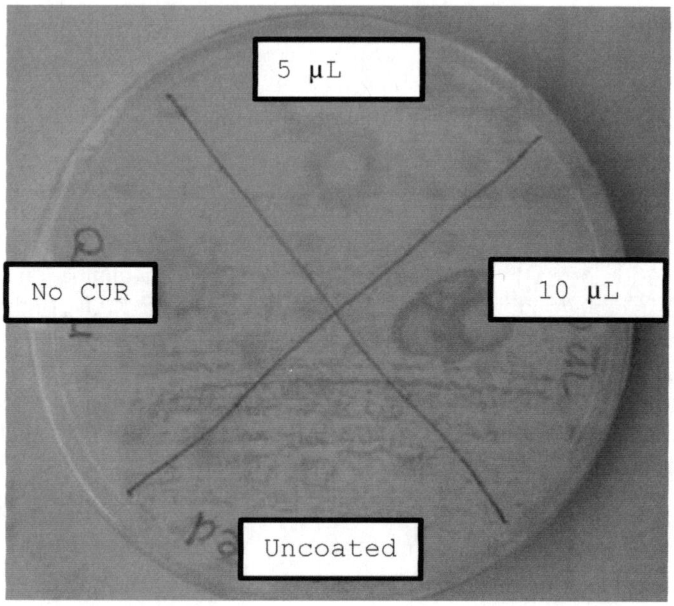

Figure 2. The inhibition zone after 3 days of incubation period.

3.3. Physical appearance

The appearance of strawberry sample was observed for five days at room temperature of 29 °C. From Figure 3, the non-coated strawberry sample showed unsatisfactory of appearance after one day of storage period as juice leakage happened on the second day of storage. The strawberry samples coated with chitosan and addition of 10 ul of turmeric oil could last up to three days. On the other hand, for the other strawberry samples, their appearances were satisfied for two days before wound, black spot and juice leakage could be seen on their surface.

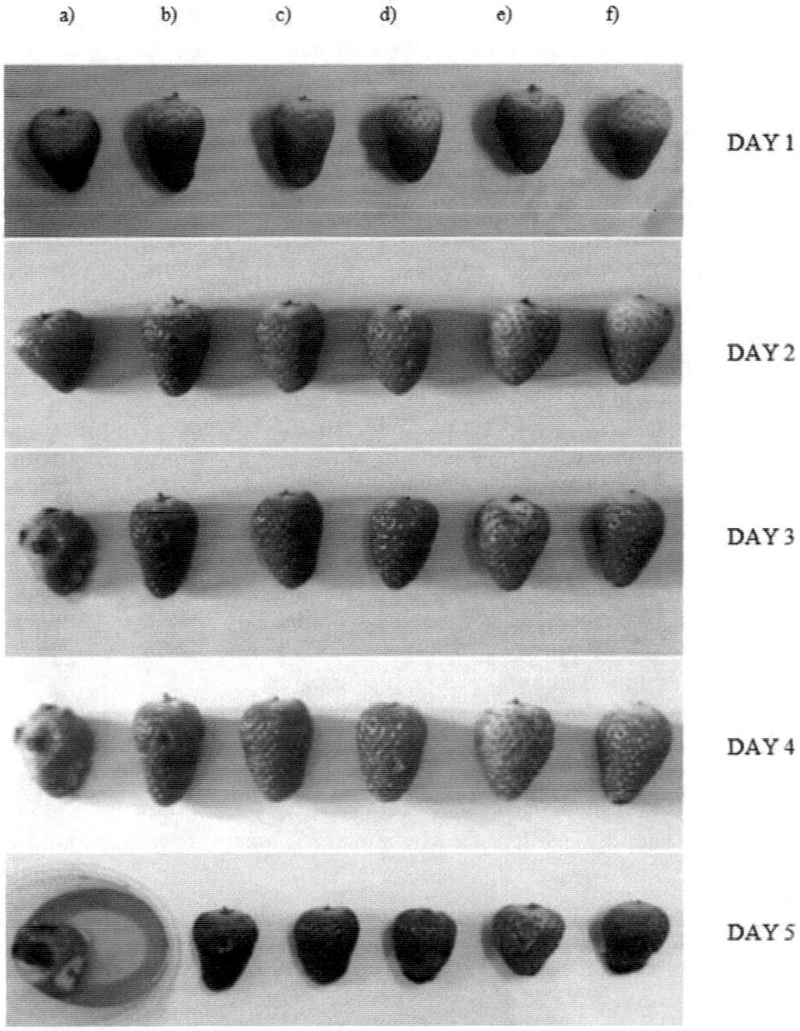

Figure 3. Appearance of control and coated strawberries at different concentrations CUR essential oil: a) control, b) 0 µL, c) 5 µL, d) 10 µL, e) 15 µL and f) 20 µL in chitosan-starch based coating.

4. Conclusion

Incorporating *Curcuma longa L.* (CUR) essential oil into chitosan-starch coating significantly affected the ripening process and reduced physicochemical changes of strawberry fruit. Chitosan-starch coating incorporated with 10 uL of CUR had the best appearance and resulted a layer of homogeneous surface about 37.45 µm as can be seen in microstructure observation. The disk diffusion analysis proves that the chitosan with 10 uL of curcuma longa had the highest inhibition zone which is 17 mm. The addition of turmeric essential oil improves the antimicrobial properties of the chitosan coating to inhibit the growth of spoilage for reducing postharvest deterioration that make them suitable as a coating for fruit and vegetable products.

IOP Conf. Series: Materials Science and Engineering **395** (2018) 012020 doi:10.1088/1757-899X/395/1/012020

5. References

[1] Guerra ICD *et al.* 2015 Coatings comprising chitosan and Mentha piperita L. or Mentha×villosa Huds essential oils to prevent common postharvest mold infections and maintain the quality of cherry tomato fruit *Int. J. Food Microbiol* **214** 168-178

[2] Xiao Z, Luo Y, and Wang Q 2011 Combined effects of sodium chlorite dip treatment and chitosan coatings on the quality of fresh-cut d'Anjou pears," *Postharvest Biol. Technol.* **62**(3) 319–326

[3] Liu K, Liu J, Li H, Yuan C, Zhong J, and Chen Y 2016 Influence of postharvest citric acid and chitosan coating treatment on ripening attributes and expression of cell wall related genes in cherimoya (Annona cherimola Mill.) fruit *Sci. Hortic* **198** 1-11

[4] Raghav PK, Agarwal N, Saini M, Vidhyapeeth J, and Vidhyapeeth J Edible Coating of Fruits and Vegetables *I nternational J. Sci. Mod. Educ* **I**(I) 188–204, 2016

[5] Kerch G 2015 Chitosan films and coatings prevent losses of fresh fruit nutritional quality: A review *Trends Food Sci. Technol.* **46** (2) Part A 159-166

[6] Xing Y *et al.* 2015 Effect of Chitosan Coating with Cinnamon Oil on the Quality and Physiological Attributes of China Jujube Fruits," *Biomed Res. Int.* **2015** 835151

[7] Martins RM, Pereira SV, Siqueira S, Salomão WF, and Freitas LAP 2013 Curcuminoid content and antioxidant activity in spray dried microparticles containing turmeric extract *Food Res. Int.* **50**(2) 657–663

[8] Maheshwari MK, Singh AK, Gaddipati J, and Srimal RC 2006 Multiple biological activities of curcumin: A short review *Life Sci.* **78**(18) 2081–2087

[9] Vásconez MB, Flores SK, Campos CA, Alvarado J, and Gerschenson LN 2009 Antimicrobial activity and physical properties of chitosan-tapioca starch based edible films and coatings *Food Res. Int.***42**(7) 762–769

[10] Santos NS *et al.* 2012 Efficacy of the application of a coating composed of chitosan and Origanum vulgare L. essential oil to control Rhizopus stolonifer and Aspergillus niger in grapes (Vitis labrusca L.) *Food Microb* **32**(2)345-53

[11] Khaliq G, Mohamed M T M, Ding P, Ghazali H M, and Ali A 2016 Storage behaviour and quality responses of mango (Mangifera indica L.) fruit treated with chitosan and gum arabic coatings during cold storage conditions *Int. Food Res. J* **23**(Suppl) S141-S148

[12] Guerreiro A C, Gago C M L, Faleiro M L, Miguel M G C, and Antunes M D C 2015 Raspberry fresh fruit quality as affected by pectin- and alginate-based edible coatings enriched with essential oils *Sci. Hortic* **194** 138-146

[13] Perdones A, Sánchez-González L, Chiralt A, and Vargas M 2012 Effect of chitosan-lemon essential oil coatings on storage-keeping quality of strawberry *Postharvest Biol. Technol.* **70** 32-41

[14] Egan M. E *et al.* 2004 Curcumin , a Major Constituent of Turmeric , Corrects Cystic Fibrosis Defects *Science* **304**(5670) 600-2

[15] Jayaprakasha G K, Rao M and Sakariah K K 2005 Chemistry and biological activities of C . longa *Trends in Food Science & Technology* **16** 533–548,

Acknowledgement

The authors would like to acknowledge Universiti Teknologi MARA for the financial support to carry out this research through 600-IRMI/MYRA 5/3/LESTARI (0047/2016). Authors also thankful to the staff of Faculty of Chemical Engineering, Universiti Teknologi MARA, Shah Alam for their help in performing laboratory work.

Metal Oxide based Antibacterial Membrane

M Zunita*, I G B N Makertihartha, F A Saputra, Y S Syaifi, and I G Wenten

Department of Chemical Engineering, Institut Teknologi Bandung, Ganesha Street 10, 40132, Bandung, West Java, Indonesia

* Email: m.zunita@che.itb.ac.id

Abstract. Biofouling is one of the major problems that hamper the application of membrane processes. Biofouling causes performance deterioration and higher energy consumption due to membrane pore blockage. In addition, it also results in a higher cleaning frequency and a shorter membrane lifetime which increases membrane operating and maintenance cost. Preparation or modification of membrane with antibacterial properties has been considered as one of the attractive strategies to control biofouling. Recently, the combination of polymeric with biocidal materials has attracted a great attention in the preparation of antibacterial membrane. A lot of metal oxide nanoparticles which have biocidal characteristics have been used to prepare nanocomposite membrane with remarkable antibacterial properties and significant biofouling reduction. According to studies that have been reported, the enhancement of anti-biofouling of the membrane was due to an interaction between the metal oxides and microorganism on the membrane surface which prevents biofilm formation. The enhancement of metal oxide nanoparticles antibacterial effect has been studied, including the elaboration of nanostructured oxide consisting of two or metallic components. In this paper, the metal oxide-based antibacterial membrane is comprehensively reviewed. In addition, mechanism of metal oxide particles in preventing biofilm formation will be discussed.

Keywords: antibacterial membrane, biofouling, biofilm, metal oxide, nanoparticle

1. Introduction

Several technologies have been used for water and wastewater treatment, such as filtration, sedimentation, flocculation, coagulation, activated carbon adsorption, and biologically active sludge technology[1, 2]. Among all of water and wastewater treatments, membrane processes are more favorable as they are cheap, fast, highly selective, and flexible to be integrated with other processes [3-10]. Although there are so many major benefits of membranes over conventional technologies, their worst facing problem is biofouling [11]. The main consequence of fouling problem is blockage of the pores of membranes, which contributes to the tremendous decrease of membrane flux and more energy necessity for filtration [12-15]. Therefore, low biofouling membranes are of major interests in numerous fields including water and wastewater treatment or food industry[16]. This could be achieved by changing membrane's chemical structure as bulk and surface modification or using nanoparticles through the membrane matrix or on the surface [17-21].

Several types of nanoparticles and their derivatives like nanometals have given high attention for some researchers because of their potential antibacterial effects to remove biofilm on the membrane surface [22]. Furthermore, morphological and physicochemical characteristics of the nanometals have been proven to exert an effect on their antimicrobial activities[23]. The exact mechanisms for the antibacterial effect of nanometals are still being investigated. But, there are two proposed possibilities those are free metal ion toxicity arising from the dissolution of the metals from surface of the nanoparticles and oxidative stress via the generation of reactive oxygen species (ROS) on surfaces of the nanoparticles[24]. In this article, we focused on the latest findings of antimicrobial activity of membranes with metal oxides as nanoparticles for wastewater treatment.

2. Characteristic of Metal Oxide as Antibacterial Agent

In general biofilm formation consist three stages: planktonic and initial interaction stages, development and maturation stages, bacteria detachment stages[25]. Biofilm formation can be stopped by blocking of the initial interaction stages from bacteria toward membrane surface. The attachment and development of biofilms on membrane surfaces is usually controlled by physical washing, such as air scrubbing, crossflow filtration, and backwashing and or using chemical substance such as dispersants and biocides[26, 27]. One of biocides that can be used to prevent the growth of biofilm is metal oxide nanoparticles. This addition of metal oxide on membrane surface aims to prevent initial interaction and development stages of biofilm formation. There are three synthesis methods of metal oxide nanoparticles, namely solution-based, vapor based, and biological based. The choice of synthesis methods will determine physicochemical properties of metal oxide nanoparticles[28]. Many metal oxide nanoparticles have been reported to have a remarkable antibacterial property, such as titanium oxide (TiO_2), silver oxide (Ag_2O), silica oxide (SiO_2), cuprum oxide (CuO), and much more[28, 29].

Many studies have reported how the metal oxides can have biocidal activity. But, there are two more popular proposed antibacterial mechanism of metal oxide nanoparticles: 1) Free metal ion toxicity arising from the dissolution of the metals from surface of the nanoparticles, 2) Oxidative stress via the generation of reactive oxygens species (ROS) on the surfaces of nanoparticles[24]. The antibacterial properties of metal oxide nanoparticles were affected by morphological and physicochemical properties of nanomaterials[22, 30]. Size and shape of metal oxide nanoparticles also affected its antibacterial activity[29]. Table 1 shows the antibacterial activity of each nanoparticle and its physical properties that affect their antibacterial activity.

Table 1. Antibacterial mechanism of several metal oxide nanoparticles.

Nanoparticles	Mechanism	Characteristic	Influence factor	Ref.
Ag_2O	DNA loses its replication ability and the cell cycle halts at the G2/M phase owing to the DNA damage	High antimicrobial activity against both bacteria and drug-resistant bacteria, antifungal activity on spore-producing fungal plant pathogens, high stability, nontoxicity.	Particle size and shape of particles	[31]
ZnO	ROS generation on the surface of the particles; zinc ion release, membrane dysfunction; and nanoparticles internalization into cell.	Photocatalytic activity; high stability; bactericidal effects on both Gram-positive and Gram-negative bacteria; antibacterial activity against spores which are resistant to high temperature and high pressure	Particle size and concentration	[23, 32-36]
TiO2	Oxidative stress via the generation of ROS; lipid peroxidation that cause to	Suitable photocatalytic properties; high stability; effective antifungal for	Crystal structure, shape and size	[35, 37-40]

	enhance membrane fluidity and disrupt the cell integrity.	fluconazole resistant strains.		
CuO	Crossing of nanoparticles from the bacteria cell membrane and then damaging the vital enzymes of bacteria.	Effective against Gram-positive and Gram-negative bacteria; high stability; antifungal activity.	Particle size and concentration	[34, 41, 42]
MgO and CaO	Damaging the cell membrane and then causing the leakage of intracellular contents and death of the bacterial cells.	Effective against both Gram-positive and Gram-negative bacteria; high stability; low cost; availability.	Particle size, pH and concentration.	[43-47]

3. Modification Membrane with Metal Oxide Nanoparticles

Biofouling is the result of interactions of biomass (including bacterial cells, cell debris, extracellular polymeric substances (EPS, etc.) with the membranes, so that it makes biofilms formation or accumulation of microbial products on the membrane surfaces even in the membrane pores[26]. To prevent this biofouling, we need to modify the membrane using the antibacterial agent, like metal oxide. Many researches show the modified membrane using a metal oxide that gives the membrane antibacterial properties and also improved the membrane properties. Nanoparticles play two critical roles in membrane forming: 1) acting as structurally strong members with greater compact and fouling resistance and 2) acting as materials to produce desirable membrane morphology [41, 48-50]. One of the common methods to synthesize metal oxide doped membranes is to use well-controlled and homogenized metal oxide particles as additives to the casting solution prior to synthesis procedures such as phase separation/immersion precipitation[51]. Metal oxides represent an important class of materials used for developing matrix membranes with enhanced properties in comparison with the purely polymeric membranes[52]. The result of polymeric membranes using metal oxides additives can be shown in Table 2.

Table 2. Features of metal oxide based antibacterial membrane.

Matrix	Nanoparticles	Result	Ref.
Polyvinylidene fluoride (PVDF)	TiO_2	High antifouling properties, increased photocalytic properties	[31]
	Al_2O_3	Enhanced antibiofouling ability and flux recovery ratio	[23, 32-36]
Polyether sulfone (PES)	TiO_2, Al_2O_3 and ZrO_2	Higher affinity, reduced membrane fouling resistance, increased hydrophilicity and rejection	[23, 32-36]
Polyvinyl chloride (PVC)	TiO_2	Greater water flux, enhanced antifouling properties, increased hydrophilicity	[35, 37-40]
	ZnO	Lower contact angle, higher hydrophilicity, greater water flux	[34, 41, 42]
Polysulfone (PSF)	TiO_2	Enhanced mechanical strength, hydrophilicity, antifouling ability and permeability	[34, 41, 42]
	SiO_2	Improved water flux, protein rejection and antifouling tendency	[34, 41, 42]
	ZnO	Enhanced water flux, hydrophilicity, permeability, porosity, rejection tendency and fouling propensity	[34, 41, 42]

As we know from Tabel 2, that all modified polymeric membranes have improved its performance from its pristine membranes. Some result using polysulfone membrane which added with Zinc Oxide, Titanium Oxide, and Silica Oxide can be seen in Table 3 respectively.

Table 3. Filtration Properties of Pristine PSf Membrane and Modified PSf membrane with Metal Oxides[53-55].

Parameter	Pure PSf	PSf-ZnO	PSf-ZnO-GO	PSf-TiO_2	PSf-TiO_2-GO	PSf-SiO_2	PSf-SiO_2-GO
Concentration (% wt)	-	2	0.6	0.5	0.5	0.3	0.3
Porosity (%)	75.2	86.7	90	N/A	N/A	N/A	N/A
Mean RMS (nm)	5.09	3.45	4.03	27.02	22.53	N/A	N/A
Contact angle (degree)	65.9	39.6	38.5	70.51	68.39	64	63
Permeability (L/m^2hbar)	0.89	2.83	5.11	0.55	0.58	N/A	N/A
Pure water flux (L/m^2h)	11	35	52	140.52	297.65	241	379
Flux recovery ratio (%)	85	97	99	N/A	N/A	69	72

From Table 3 shows that the addition of metal oxide nanoparticles successfully improve the performance of polysulfone membrane. This result also shows, that metal oxide not only gives membrane antibacterial properties but also increase filtration characteristic of the membrane. Antibacterial activity from the modified polymeric membrane using metal oxide can be observed using Scanning Electron Microscope (SEM).

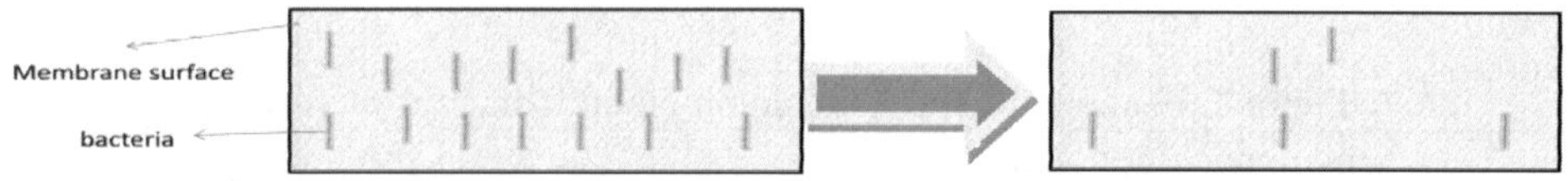

Figure 1. The illustration of antibacterial activity on the membrane surface. Adopted from[53, 56]

Based on Figure 1 (the illustration of antibacterial activity on the membrane surface) shows that the addition of metal oxide can decrease microorganisms attachment on the membrane surfaces. The metal oxide on the polysulfone membrane surface can remove almost all bacteria colonies by using ZnO-GO and Ag-GO antibacterial agents. Other than that, the addition of Ag-SiO_2 and Ag@TiO_2-CNTs make the bacteria colonies die. The antibacterial activity of metal oxide will vary depending on its antibacterial mechanism, from Figure 1 emphasizes that the number of bacterial colonies on the membrane surface due to the amount or type of different metal oxides.

4. Challenge on Preparation of Metal Oxide Based Antibacterial Membrane

To improve membrane performance, many modifications have been made to the membrane preparation [51, 57-60]. Membrane preparation with the addition of metal oxide can increase membrane antibacterial, antifouling, and membrane filtration properties. Antibacterial mechanisms in the metal oxide involve the release of metal ions into the water which causes the amount of metal oxide in the membrane surface decreases [22]. The improvement of interaction between metal oxide and membrane surface is required to reduce the rate of metal ions released from the membrane surface so that maintain the antibacterial activity can go longtime, moreover can reduce soluble metal ions concentration in water. From Table 5 shows that combining two metal oxides or with another metal nanoparticle can increase membrane performance. There many studies reported that optimized metal nanoparticles exhibited better antibacterial properties than a single metal oxide [61-65]. The other challenge is to combine the metal oxide with another natural resource, to give better antibacterial activity and friendly to the environment. Some research has shown the combination from ZnO and Eugenol to give antibacterial activities in polysulfone membrane [66]. Next challenge to improve the antibacterial property of membrane is combine various polymeric membranes with nanoparticles to find the most appropriate combinations and applications of the membrane fabricated. For example, in drinking water application, metal oxide must handle in carefully due to potentially its toxicity toward the human. Moreover, the antibacterial agent activity that only prevents microorganism growth on membrane surface. Beside of that, the polymeric materials and nanoparticles researches need to be optimized so that cost of production more competitive, give high performance and longtime antibacterial activity of membranes.

5. Conclusion

This current review is aimed to discuss about the utilization of metal oxide nanoparticles in membrane to prevent biofouling activity in the membrane surface. Based on these and earlier studies, Membrane is an advanced treatment technology that has many benefits such as cheap, highly selective, using less chemicals and energy, simple design, and easy to maintenance. Membrane also shows great performance for wastewater treatment application in many industries by using physical filtration processes or membrane bioreactor. In other hand, biofouling causes the blockage of the membranes pores and reduces the membrane performance. But, it can be removed by modifying the membrane with antibacterial agent like metal oxide nanoparticles. The proposed antibacterial mechanism of metal oxide nanoparticles are free metal ion toxicity from dissolution of metal oxide nanoparticles and oxidative stress via the generation of ROS on the surface of nanoparticles. The antibacterial properties of metal oxide nanoparticles was affected by morphological and physicochemical properties of nanomaterials such as crystal structure, shape, size, concentration, and pH. The modified membrane using a metal oxide nanoparticles has better antifouling properties and also improved other properties such as mechanical strength, water flux, hydrophilicity, permeability, porosity, and rejection tendency. It is expected that this review may be able to enhance further research into the development of metal oxide nanoparticles in membrane technology for many application, especially wastewater treatment.

6. References

[1] Metcalf, E. and E. Eddy 2003 *Wastewater engineering: treatment and reuse. McGrawHill.* Inc., New York

[2] Krantz, D. and B. Kifferstein 2005 Water pollution and society *Journal of Environmental Law Practice* **4**(1) 1-2

[3] Juang, L.-C., D.-H. Tseng, and H.-Y. Lin 2007 Membrane processes for water reuse from the effluent of industrial park wastewater treatment plant: a study on flux and fouling of membrane *Desalination* **202**(1-3) 302-309

[4] Wenten, I. and Khoiruddin 2016 Reverse osmosis applications: prospect and challenges *Desalination* **391** 112-125

[5] Wenten, I.G., H. Julian, and N.T. Panjaitan 2012 Ozonation through ceramic membrane contactor for iodide oxidation during iodine recovery from brine water *Desalination* **306** 29-34

[6] Wenten, I.G. and F. Arfianto 2013 Bench scale electrodeionization for high pressure boiler feed water Desalination **314** 109-114

[7] Khoiruddin, K., A. Hakim, and I.G. Wenten 2014 Advances in electrodeionization technology for ionic separation-A review *Membrane Water Treatment* **5** (2) 87-108

[8] Wenten, I., et al. 2017 LTA zeolite membranes: current progress and challenges in pervaporation *RSC Advances* **7**(47) 29520-29539

[9] Wardani, A.K. Hakim, A.N., Khoiruddin, Wenten, I.G. 2017 Combined ultrafiltration-electrodeionization technique for production of high purity water *Water Science and Technology* **75**(12) p. 2891-2899

[10] Ariono, D. and I.G. Wenten 2017 Surface modification of ion-exchange membranes: Methods, characteristics, and performance *J. Appl. Polym. Sci* **134**, 45540

[11] Guo, W., H.-H. Ngo, and J. Li 2012 A mini-review on membrane fouling *Bioresource technology* **122** p. 27-34

[12] Asatekin, A., et al., 2007 Anti-fouling ultrafiltration membranes containing polyacrylonitrile-graft-poly (ethylene oxide) comb copolymer additives *Journal of Membrane Science* **298**(1): p. 136-146

[13] Wenten, I. G 1995 Mechanisms and control of fouling in crossflow microfiltration *Filtration & separation* **32**(3) p. 252-253

[14] Wenten, I. G. and Widiasa, I.N. 2002 Enzymatic hollow fiber membrane bioreactor for penicilin hydrolysis *Desalination* **149**(1) 279-285

[15] Chen, V., et al. 1997 Particle deposition during membrane filtration of colloids: transition between concentration polarization and cake formation *Journal of Membrane Science* **125**(1) 109-122

[16] Chisti, Y 2007 Biodiesel from microalgae *Biotechnology advances* **25**(3) 294-306

[17] Chung, T.-S., et al., 2007 Mixed matrix membranes (MMMs) comprising organic polymers with dispersed inorganic fillers for gas separation *Progress in Polymer Science* **32**(4) 483-507

[18] P.T.P. Aryanti, R. Yustiana, R.E.D. Purnama, and I.G. Wenten 2015 Performance and characterization of PEG400 modified PVC ultrafiltration membrane *Membrane Water Treatment* **6**(5)

[19] P.T.P. Aryanti, S. R. Joscarita, A. K. Wardani, S. Subagjo, D. Ariono, and I. G. Wenten 2016 The Influence of PEG400 and Acetone on Polysulfone Membrane Morphology and Fouling Behaviour. *Journal of Engineering and Technological Sciences* **48**(2) 135-149

[20] Himma, N.F., A.K. Wardani, and I.G. Wenten 2017 Preparation of Superhydrophobic Polypropylene Membrane Using Dip-Coating Method: The Effects of Solution and Process Parameters *Polymer-Plastics Technology and Engineering* **56**(2) 184-194

[21] Ariono, D. and I. Wenten 2017 Heterogeneous structure and its effect on properties and electrochemical behavior of ion-exchange membrane *Materials Research Express* **4**(2) 024006

[22] Seil, J.T. and T.J. Webster 2012 Antimicrobial applications of nanotechnology: methods and literature *Int J Nanomedicine* **7** 2767-2781

[23] Ravishankar Rai, V. and A. Jamuna Bai 2011 *Nanoparticles and their potential application as antimicrobials.* Science against microbial pathogens: communicating current research and technological advances. Badajoz, Spain: Formatex Research Center, p. 197-209

[24] Besinis, A., T. De Peralta, and R.D. Handy 2014 The antibacterial effects of silver, titanium dioxide and silica dioxide nanoparticles compared to the dental disinfectant chlorhexidine on Streptococcus mutans using a suite of bioassays *Nanotoxicology* **8**(1) 1-16

[25] O'Toole, G., H.B. Kaplan, and R. Kolter 2000 Biofilm formation as microbial development *Annual Reviews in Microbiology* **54**(1) 49-79

[26] Miura, Y., Y. Watanabe, and S. Okabe 2007 Membrane biofouling in pilot-scale membrane bioreactors (MBRs) treating municipal wastewater: impact of biofilm formation *Environmental science & technology* **41**(2) 632-638

[27] Ridgway, H.F. and H.-C. Flemming, 1996 *Membrane biofouling in water treatment membrane processes.* McGraw-Hill, Columbus, OH

[28] Stankic, S., et al., 2016 Pure and multi metal oxide nanoparticles: synthesis, antibacterial and cytotoxic properties *Journal of Nanobiotechnology* **14**(1) 73

[29] Dizaj, S.M., et al., 2014 Antimicrobial activity of the metals and metal oxide nanoparticles *Materials Science and Engineering: C* **44** 278-284

[30] Mohammadi, G., et al., 2010 Development of azithromycin–PLGA nanoparticles: Physicochemical characterization and antibacterial effect against Salmonella typhi *Colloids and Surfaces B: Biointerfaces* **80**(1) 34-39

[31] Sondi, I. and B. Salopek-Sondi 2004 Silver nanoparticles as antimicrobial agent: a case study on E. coli as a model for Gram-negative bacteria *Journal of colloid and interface science* **275**(1) 177-182

[32] Saraf, R 2013 Cost effective and monodispersed zinc oxide nanoparticles synthesis and their characterization *International Journal of Advances in Applied Sciences* **2**(2) 85-88

[33] Liu, Q., et al., 2014 Effects of ZnO nanoparticles and microwave heating on the sterilization and product quality of vacuum-packaged Caixin. *Journal of the Science of Food and Agriculture* **94**(12) 2547-2554

[34] Azam, A., et al., 2012 Antimicrobial activity of metal oxide nanoparticles against Gram-positive and Gram-negative bacteria: a comparative study *International journal of nanomedicine* **7** 6003

[35] Rai, M., A. Yadav, and N. Cioffi, 2012 Silver nanoparticles as nano-antimicrobials: bioactivity, benefits and bottlenecks, in *Nano-Antimicrobials*., Springer. p. 211-224

[36] Rasmussen, J.W., et al., 2010 Zinc oxide nanoparticles for selective destruction of tumor cells and potential for drug delivery applications *Expert opinion on drug delivery* **7**(9) 1063-1077

[37] Allahverdiyev, A.M., et al., 2011 ntimicrobial effects of TiO2 and Ag2O nanoparticles against drug-resistant bacteria and leishmania parasites *Future microbiology* **6**(8) 933-940

[38] Haghighi, F., et al., 2013 Antifungal activity of TiO2 nanoparticles and EDTA on Candida albicans biofilms *Infection, Epidemiology and Medicine* **1**(1) 33-38

[39] Roy, A.S., et al., 2010 Effect of nano-titanium dioxide with different antibiotics against methicillin-resistant Staphylococcus aureus *Journal of Biomaterials and Nanobiotechnology* **1**(01) 37

[40] Carré, G., et al., 2014 TiO2 photocatalysis damages lipids and proteins in Escherichia coli. *Applied and environmental microbiology* **80**(8) 2573-2581

[41] Ahamed, M., et al., 2014 Synthesis, characterization, and antimicrobial activity of copper oxide nanoparticles *Journal of Nanomaterials* **2014** 17

[42] Mahapatra, O., et al., 2008 Ultrafine dispersed CuO nanoparticles and their antibacterial activity *Journal of Experimental Nanoscience* **3**(3) 185-193

[43] Yamamoto, O., et al., 2010 Antibacterial characteristics of CaCO 3–MgO composites *Materials Science and Engineering: B* **173**(1) 208-212

[44] Jin, T. and Y. He 2011 Antibacterial activities of magnesium oxide (MgO) nanoparticles against foodborne pathogens *Journal of Nanoparticle Research* **13**(12) 6877-6885

[45] Hewitt, C.J., et al., 2001 An evaluation of the anti-bacterial action of ceramic powder slurries using multi-parameter flow cytometry *Biotechnology Letters* **23**(9) 667-675

[46] Leung, Y.H., et al., 2014 Mechanisms of Antibacterial Activity of MgO: Non-ROS Mediated Toxicity of MgO Nanoparticles Towards Escherichia coli *Small* **10**(6) 1171-1183

[47] Vidic, J., et al., 2013 Selective antibacterial effects of mixed ZnMgO nanoparticles *Journal of Nanoparticle Research* **15**(5) 1595

[48] Reyes, V.C., S.O. Opot, and S. Mahendra 2015 Planktonic and biofilm-grown nitrogen-cycling bacteria exhibit different susceptibilities to copper nanoparticles *Environmental Toxicology and Chemistry* **34**(4) 887-897

IOP Conf. Series: Materials Science and Engineering **395** (2018) 012021 doi:10.1088/1757-899X/395/1/012021

[49] Bottino, A., et al., 2001 *Preparation and properties of novel organic–inorganic porous membranes.* Separation and Purification Technology **22** 269-275
[50] Choi, J.-H., J. Jegal, and W.-N. Kim 2006 Fabrication and characterization of multi-walled carbon nanotubes/polymer blend membranes *Journal of Membrane Science* **284**(1) 406-415
[51] Arsuaga, J.M., et al., 2013 Influence of the type, size, and distribution of metal oxide particles on the properties of nanocomposite ultrafiltration membranes *Journal of membrane science* **428** 131-141
[52] Kim, J. and B. Van der Bruggen 2010 The use of nanoparticles in polymeric and ceramic membrane structures: review of manufacturing procedures and performance improvement for water treatment *Environmental Pollution* **158**(7) 2335-2349
[53] Chung, Y.T., et al., 2017 Development of polysulfone-nanohybrid membranes using ZnO-GO composite for enhanced antifouling and antibacterial control *Desalination* **402** 123-132
[54] Sirinupong, T., et al., 2017 Synthesis and characterization of thin film composite membranes made of PSF-TiO 2/GO nanocomposite substrate for forward osmosis applications *Arabian Journal of Chemistry* (In press)
[55] Wu, H., B. Tang, and P. Wu 2015 Development of novel SiO 2–GO nanohybrid/polysulfone membrane with enhanced performance. Journal of Membrane Science, 2014. **451**: p. 94-102.
[56] Mahmoudi, E., et al., Novel nanohybrid polysulfone membrane embedded with silver nanoparticles on graphene oxide nanoplates *Chemical Engineering Journal* **277** 1-10
[57] Tan, K. and S.K. 2007 Obendorf Development of an antimicrobial microporous polyurethane membrane *Journal of Membrane Science* **289**(1) 199-209
[58] Tan, K. and S.K. Obendorf 2007 Fabrication and evaluation of electrospun nanofibrous antimicrobial nylon 6 membranes *Journal of Membrane Science* **305**(1) 287-298
[59] Yao, F., et al., 2008 Antibacterial effect of surface-functionalized polypropylene hollow fiber membrane from surface-initiated atom transfer radical polymerization. *Journal of Membrane Science* **319**(1) 149-157
[60] Isloor, A.M., et al., 2013 Studies on copper coated polysulfone/modified poly isobutylene alt-maleic anhydride blend membrane and its antibiofouling property *Desalination* **308** 82-88
[61] Liu, W., et al., 2014 Synthesis of TiO 2 nanotubes with ZnO nanoparticles to achieve antibacterial properties and stem cell compatibility *Nanoscale* **6**(15) 9050-9062
[62] Gordon, T., et al., 2011 Synthesis and characterization of zinc/iron oxide composite nanoparticles and their antibacterial properties *Colloids and Surfaces A: Physicochemical and Engineering Aspects* **374**(1) 1-8
[63] Zawadzka, K., et al., 2016 Mechanisms of antibacterial activity and stability of silver nanoparticles grown on magnetron sputtered TiO2 coatings *Bulletin of Materials Science* **39**(1) 57-68
[64] Sharma, N., et al., 2016 Synthesis, characterisation and antimicrobial activity of manganese-and iron-doped zinc oxide nanoparticles *Journal of Experimental Nanoscience* **11**(1) 54-71
[65] Chanhom, P., et al., 2016 Colloidal titania-silica-iron oxide nanocomposites and the effect from silica thickness on the photocatalytic and bactericidal activities *Journal of Magnetism and Magnetic Materials* **427** 54-59
[66] Harun, Z., et al. 2016 Antibacterial Polysulfone Membranes: The Effect of Eugenol and Zinc Oxide as Additives *Materials Science Forum* **867** 132-138

The Study of Mechanical Properties and Relaxation Time of Agar Hydrogel for Tissue Mimicking Phantom Material in Magnetic Resonance Imaging

Y Dwihapsari, R D Afriana, Darminto

Department of Physics, Institut Teknologi Sepuluh Nopember Surabaya (ITS), Kampus ITS Sukolilo, Surabaya 60111 Indonesia

yanuritadh@physics.its.ac.id, yanuritadh@gmail.com

Abstract. Agar hydrogel is used for tissue mimicking phantom in magnetic resonance imaging studies. The spin-lattice relaxation time, T_1 as an important parameter in Magnetic Resonance Imaging measurement was observed. The T_1-value represents how fast the net spin magnetization vector recovers to its equilibrium after generation of magnetic field. To explore mechanical properties of agar, shear modulus was examined using a dynamic mechanical analyser. The relation between mechanical properties and spin-lattice relaxation time of agar hydrogel is discussed.

Keywords: hydrogel, agar, magnetic resonance imaging, mechanical properties, phantom

1. Introduction

Magnetic Resonance Imaging (MRI) uses magnetic nature of atomic nuclei for observing and analysing physical and chemical properties of molecules inside a material. When the material is placed inside the magnetic field, the spin of nuclei will align with the field which gives rise to a macroscopic magnetisation and precess with Larmor frequency. The energy transfer between ground state and higher energy level takes place following this rate and energy is emitted when spin returns to its ground state. The signal is detected when magnetisation induces a current in the coil via Faraday induction. It is then processed and reconstructed to be displayed as a spectrum or image.

In MRI study, the properties of material are observed using some parameters, such as spin-lattice relaxation time (T_1) and spin-spin relaxation time (T_2). The T_1 refers to time needed for spins to relax and return toward thermal equilibrium after interaction of spins with magnetic field. T_1-value is related with energy dissipated to surrounding environment or lattice and depends on the nature and rate of molecular motion inside the material.

Agar as a tissue-mimicking material for MRI study had been discussed in some studies because of its low cost and easiness to adjust with spin-lattice relaxation rates of human tissues. Several studies had observed T_1 properties of hydrogels and found a significant relation between T_1-value and hydrogel concentrations [1, 2]. Some studies found that T_1 decreases linearly with the increasing agar concentration [1]. Although lower T_1 can be achieved by increasing agar concentration, to some extent it is still higher compared to that of some tissues. Therefore, the paramagnetic agent such as $CuSO_4$ is generally used to further decrease T_1 to match the one of corresponding tissues. The previous study showed that T_1 of $CuSO_4$ decreased as its concentration increased [1].

In the studies of tissue mimicking phantom, the mechanical properties of agar are observed to determine the compressive force, so that agar could withstand while retaining their shapes. The mechanical properties, such as elastic, viscous and shear modulus were found to be higher as agar concentration increases [3]. In this study, shear modulus which describes the response of material to shear stress is examined.

The effect of agar concentration on shear modulus and spin-lattice relaxation time T_1 will be discussed in this study. Agar with various concentrations were measured ca. 2 h after gelation and scanned using MRI to obtain T_1. The shear modulus was observed using dynamic mechanical analyser (DMA). The relation between shear modulus and T_1 is plotted to explore their link to agar concentration.

2. Materials and methods

Agar powder purchased from local market was mixed with distilled water to form gel. Agar samples were prepared with various concentration of agar mixed with 50 ml distilled water and boiled for 20 min to form 2.5, 5, and 7.5 w/v% of agar gel. The mixture was poured into plastic tube of 57-mm diameter to fill 75 mm height and used without further purification. Agar gel was solidified by putting it in room temperature and scanned after 2 – 4 h of preparation.

The relaxation time measurement was conducted using 1.5 T MRI scanner (Signa, General Electric, US) with spine gradient coil. Samples were arranged according to concentration of agar and $CuSO_4$. Spin-lattice relaxation rate measurement was conducted using inversion recovery spin-echo sequences explained in Hellerbach *et.al* [4]. In order to assess the homogeneity of sample, four slices were taken from centre of gradient coil in each sample with slice thickness of 5 mm. The parameters used for this study are: repetition time (TR) = 500 – 10000 ms, echo-time (TE) = 10 – 150 ms, field-of view (FOV) = 40 mm × 40 mm, and matrix size = 64 × 64. The images were then analysed using ImageJ to obtain signal intensity and T_1 was obtained by fitting the mean signal intensity into TR.

The shear modulus of agar was observed using DMA Mettler Toledo SDTA861. The stress-strain relation curve was plotted to obtain the shear modulus. To explore the relationship between relaxation rates and mechanical properties of agar, spin-lattice relaxation time was plotted against shear modulus.

3. Results and discussion

Table 1 shows T_1 relaxation time of various concentration and mixture of agar and $CuSO_4$, showing T_1 which decreases as agar concentration increases. The addition of paramagnetic agent such as $CuSO_4$ gives lower T_1 and shows the relaxation time which is dependent on $CuSO_4$ concentration. T_1-value of agar and agar with $CuSO_4$ are in a good agreement with those in other studies [1] and T_1 of human tissue ranging between 500 – 1500 ms at 1.5 T [5]. The lower agar concentration allows more free water protons and results in longer T_1. Contrarily, higher concentration of agar gives more barrier and collisions to protons and results in shorter T_1. The paramagnetic agent $CuSO_4$, gives even lower T_1 and make it a potential material for production of low-T_1 tissue-mimicking phantom.

Figure 1 describes the shear modulus of agar hydrogel and shows that shear modulus increases as agar concentration increases. The shear modulus from very low agar concentration cannot be obtained due to broken sample during testing thus sample of 2.5% agar concentration was excluded in our study. The result of shear modulus is consistent with the previous study that showed higher shear modulus due to higher agar concentration [6]. The shear modulus of agar ranges between 130 - 180 kPa for 5% and 300 – 380 kPa for 7.5% of agar concentration. The shear modulus is not significantly affected by addition of $CuSO_4$ as shown in Fig.1 although higher $CuSO_4$ concentration increases the elastic region of agar especially for 5.9% $CuSO_4$.

Table 1. Spin-lattice relaxation time, T_1 of various concentration of agar and mixture of agar and $CuSO_4$.

Sample	Agar concentration (w/v%)	$CuSO_4$ concentration (w/v%)	T_1 (ms)	SD
1	2.5	0	1794.713	122.187
2	5	0	1540.585	220.167
3	7.5	0	1331.503	134.376
4	2.5	0.4	1233.339	182.094
5	5	0.4	1055.102	151.318
6	7.5	0.4	840.336	92.270
7	2.5	5.9	1073.836	196.888
8	5	5.9	869.565	58.776
9	7.5	5.9	740.741	87.221

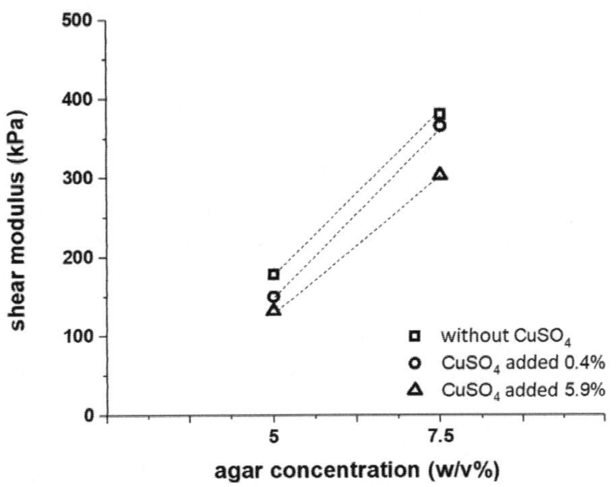

Figure 1. Shear modulus of agar and $CuSO_4$. Variation of agar concentration = 5% and 7.5%. Variation of $CuSO_4$ concentration = 0%, 0.4%, and 5.9%. Agar 2.5% was excluded from shear modulus experiment.

The result shows that increasing agar concentration results in lower T_1 and higher shear modulus as shown in Figure 2. The plot in Fig. 2 comparatively shows linear relationship between T_1 and shear modulus in pure agar as well as mixture of agar and $CuSO_4$, although the addition of paramagnetic agent does not significantly influence shear modulus. The more variation of agar and $CuSO_4$ concentration is needed for further investigation to observe the relation between shear modulus and T_1 in order to produce tissue mimicking phantom in MRI with wider range of T_1 without neglecting the mechanical properties of material.

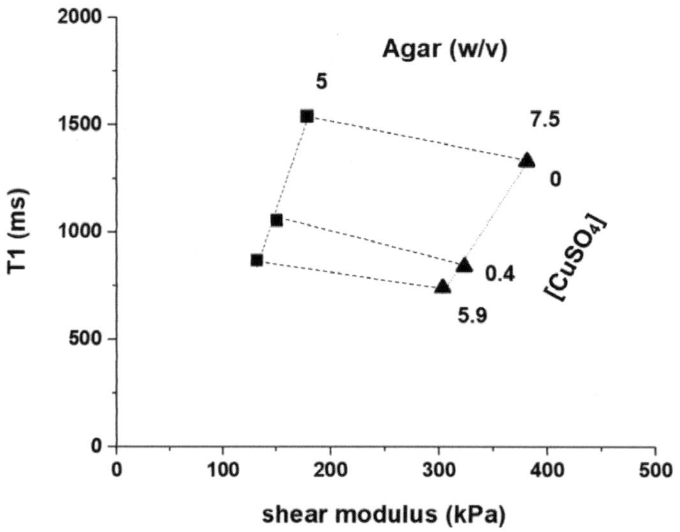

Figure 2. Shear modulus vs T_1 of agar and mixture of agar and $CuSO_4$. 1) Agar + $CuSO_4$ 0%, 2) Agar + $CuSO_4$ 0.4%, 3) Agar + $CuSO_4$ 5.9%. The result from agar 2.5% was excluded in this plot.

4. Conclusion

The study on spin-lattice relaxation time and shear modulus of agar hydrogel had been conducted. Increasing agar concentration results in lower T_1 and increasing shear modulus. $CuSO_4$ contributes to even lower T_1 for the same agar concentration but does not significantly affect shear modulus.

5. References

[1] Mitchell MD, Kundel HL, Axel L, Joseph PM 1986 *Magn Reson Imaging* **4**(3) 263-266
[2] Bucciolini M, Ciraolo L, Lehmann B 1989 *Acta Radiol.* **30**(6) 667-669
[3] Ahearne M, Yang Y, El Haj AJ, Then KY, & Liu K-K 2005 *J. of the Royal Society Interface* **2**(5) 455–463
[4] Hellerbach A, Schuster V, Jansen A, Sommer J 2013 *Plos One* **8**(8) 1 - 8
[5] Stanisz GJ et al 2005 *Magn Reson Med* **54**(3) 507-512
[6] Hamhaber U, Grieshaber FA, Nagel JH, Klose U 2003 *Magn. Res. Med* **49**(1) 71 - 77

Acknowledgments

This research is funded by Research Grant Penelitian Pemula ITS 2017. Authors wish to acknowledge Prof Suminar Pratapa and Nur Aini Fauziyah for discussion and assistance on DMA measurement.

Fabrication of Scaffold based on Chitosan – Poly-2-Acrylamido- 2-Methylpropane Sulfonic Acid (PAMPS) Polyelectrolyte Complexes

A Wibowo[1], Mumtaziah[1], L A T W Asri[1], F Qulub[1], A I Mahyuddin[2], T Dirgantara[3], R Suratman[1,a]

[1]Material Science and Engineering Research Group, Faculty of Mechanical and Aerospace Engineering, Institut Teknologi Bandung, Jl. Ganesha 10, Bandung, Indonesia
[2]Mechanical Design Research Group, Faculty of Mechanical and Aerospace Engineering, Institut Teknologi Bandung, Jl. Ganesha 10, Bandung, Indonesia
[3]Lightweight Structure Research Group, Faculty of Mechanical and Aerospace Engineering, Institut Teknologi Bandung, Jl. Ganesha 10, Bandung, Indonesia

[a] Corresponding author: rochim@material.itb.ac.id

Abstract. Chitosan is considered as one of the best biopolymers for bone tissue engineering application due to its appealing properties such as biocompatible, non-toxic, non-immunogenic, able to accelerate growth rate of bone tissue, possess antibacterial activity and able to form porous structure. Herein, polyelectrolyte complexes (PECs) prepared by simple mixing of chitosan solution and poly-2-acrylamido-2-methylpropane sulfonic acid (PAMPS) solution were investigated. In this study, chitosan concentration were varied from 1.0 to 2.0% (w/v), while concentration of AMPS was fixed at 0.1% (w/v) to know the influence of chitosan concentration on swelling capacity and mechanical properties of obtained PECs. Swelling test and compressive test results showed that the most stable PECs in aqueous environment with highest ultimate compressive strength was PECs obtained from chitosan 2.0% (w/v) – PAMPS 0.1% (w/v) with swelling capacity 3,326% for 2 hours and ultimate compressive strength 1.22 kPa.

Keywords: cancellous bone implant, chitosan, poly-2-acrylamido-2-methylpropane sulfonic acid, polyelectrolytes complexes, tissue engineering.

1. Introduction

Recently, bone fractures cases raise significantly due to higher number of road traffic accidents and bone weakening disease such as osteoporosis. Consequently, the demand of human bone implant will also increase in the following trend as the organ donor shortage.[1] However, the recent bone implant mainly focus on replacing hard and dense tissue of bone, namely cortical bone. While bone implant to replace sponge and soft tissue, namely cancellous bone, is still limited. To fulfil this demand, new materials with high porosity, similar mechanical strength ($\sigma_{u\ compressive}$ = 1 - 12 MPa)[2], induce cell-attachment and immune-acceptance is needed.

IOP Conf. Series: Materials Science and Engineering **395** (2018) 012023 doi:10.1088/1757-899X/395/1/012023

One of attractive materials to overcome this challenge is chitosan because of its appealing properties, such as biocompatible, non-toxic, non-immunogenic and has good cell attachment. However, chitosan has some disadvantages when being used for physical support because it is very stiff, brittle and low mechanical resistance ($\sigma_{u\ compressive}$ (2% w/v, dry) = 0.967 MPa).[2-5] Besides, chitosan is also unstable in the porous form that it cannot maintain its shape that will disturb the tissue regeneration and make the scaffold difficult to function as the support.

In order to optimize the mechanical properties and stability of chitosan scaffold, crosslinking agents are used to bond the polymeric chains to create a 3D network in order to enhance the mechanical properties. Even so, most of crosslinking agents are toxic to human body thus it is not safe.[6] The other way to stabilize the chitosan scaffold is pairing the cationic chitosan with negatively charge biopolymer hence they are only bonded by electrostatic interaction to form polyelectrolyte complexes (PECs) which results a non-toxic bond. The electrostatic interaction in polyelectrolyte complexes also might enhance the mechanical properties and possess better stability of chitosan. One of known negatively charged polymers is poly-(2-acrylamido-2-methyl-1-propanesulfonic acid) (PAMPS), which is a synthetic biocompatible polymers that has distinctive behaviour in complement system in order to delay the hyperacute rejection.[7] Previously, Zhang et al has been prepared chitosan-PAMPS based PECs using a polymer-monomer pair reaction system.[8] In this research, chitosan-PAMPS based PECs was prepared by simple mixing of PAMPS solution with chitosan solution. This method might offer easier and faster strategy to obtain chitosan-PAMPS based PECs, because it can be prepared directly from their polymers solution. It is expected that mechanical strength of chitosan based scaffold could be elevated by formation of PECs that is useful for bone implant application.

2. Experimental Section

2.1Materials and Raw Materials Characterizations

Chitosan was purchased from CV. Biochitosan Indonesia. Before used, chitosan was characterized by FTIR in order to determine the degree of deacetylation (DD) using baseline method.[9,10] The determination of DD is important to ensure whether purchased chitosan can be classified as chitosan with DD above 70%.[11] FTIR measurements were recorded with KBr pellets on Prestige 21 Shimadzu. The spectra were measured at wavelength 4500 to 400 cm^{-1}. After the FTIR spectra were collected, DD of sample was calculated using Eq. 2.1:

$$DD[\%] = 100 - (\frac{A_{1655}}{A_{3450}} x\ 115) \qquad (2.1)$$

2.2Preparation and Characterization of CS/PAMPS PECs

CS/PAMPS PECs were prepared by simple mixing of 0.1% (w/v) PAMPS in 150 mM sodium chloride solution and CS in 2% acetic acid solution with various concentration (1.0%, 1.5%, 2.0% (w/v)). Briefly, 50 mL CS solution was added to 50 mL PAMPS solution using dropper pipette while the solution was stirred at 300 rpm for two hours. The obtained samples were freeze dried at -50 °C overnight to obtain dried samples without destroying their structures.

Morphologies of obtained samples were characterized using SEM Hitachi SU3500. Samples were gold coated and measured at acceleration voltage between 3 – 10 kV. Mechanical properties of obtained samples were determined by performing compression test using Tensilon®. Dimension of freeze dried samples with cylindrical shape were measured by caliper. The load cell was 10 kN with test speed 1 mm/min.

Water uptake properties of obtained samples were determined by performing swelling test on four specimens of each freeze dried samples. After measuring the dry weight of test specimens (W_d), they were immersed in SBF (pH 7.1 at 37 °C) at room temperature for predetermined time intervals, and wet weights (W_w) were determined at these time intervals after removing excess surface water by tissue paper. Degree of swelling of specimens were calculated using Eq. 2.2:

$$S\,[\%] = \frac{W_w - W_d}{W_d} x100\% \quad (2.2)$$

3. Results and Discussion

3.1 FTIR result

To ensure that purchased chitosan can be classified as chitosan that was defined with their DD should be above 70%, purchased chitosan was characterized using FTIR method and FTIR spectra of sample can be seen in Fig. 1. As can be seen in Fig. 1, FTIR spectra of sample of purchased chitosan is associated to the chitosan with the characteristic intense absorption bands at ~3400 cm^{-1} that indicates presence of the vibrate stretching band of O-H functional groups and/or secondary band for N-H groups, as well as to intermolecular hydrogen bonding within the polyssacharide. The FTIR spectrum of chitosan itself showed some features of amide groups: amide I and amide II bands at 1656 and 1600 cm^{-1}, respectively. Thus, these spectrum peaks showed the distinctive functional groups in chitosan, which is secondary amide.[9,10,12,13]

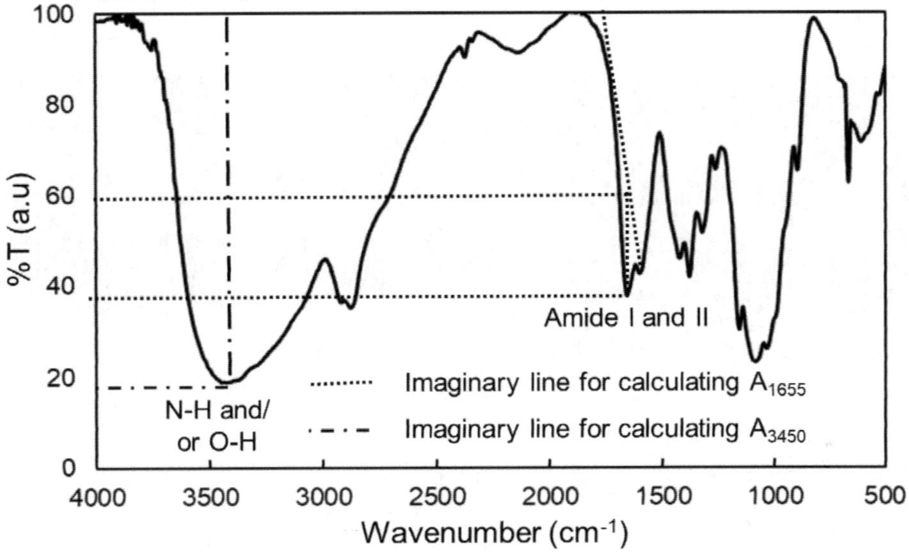

Figure 1. FTIR spectra of purchased chitosan

Based on baseline method, DD of sample was 77% using Eq. 2.1. Considering that DD of sample is above 70%, it can be claimed that purchased chitosan is classified as chitosan and can be used for this experiment. DD is above 70% means the deacetylated D-glucosamine in the sample is more dominant than the acetylated glucosamine, which is important to obtain chitosan with positive charge (polycations).

3.2 Visual and SEM observation

Visual observation of obtained freeze dried sample (Fig. 2) showed that bulkier and denser sample were observed as chitosan concentration increased. By increasing chitosan concentration, solute content increase and the amount of solvent to be sublimed decrease, lead to formation of bulkier and denser sample after freeze drying process.

IOP Conf. Series: Materials Science and Engineering **395** (2018) 012023 doi:10.1088/1757-899X/395/1/012023

Figure 2. Visual observation of obtained freeze dried sample with chitosan concentration: a) 1.0%, b) 1.5%, and c) 2.0% (scale bar represent 100 mm)

Further observation of sample morphologies were done by SEM observation (Fig.3). As can be seen in Fig. 3, sample with chitosan concentration 1.0% (Fig. 3a) showed loose structure, while sample with chitosan concentration 1.5% and 2.0% (Fig. 3b and 3c respectively) showed more compact structure. This compact structure is caused by the presence of excessive chitosan, which possesses rigid structure, that lead to the bigger remaining solute in solution that was not carried away during freeze drying process.

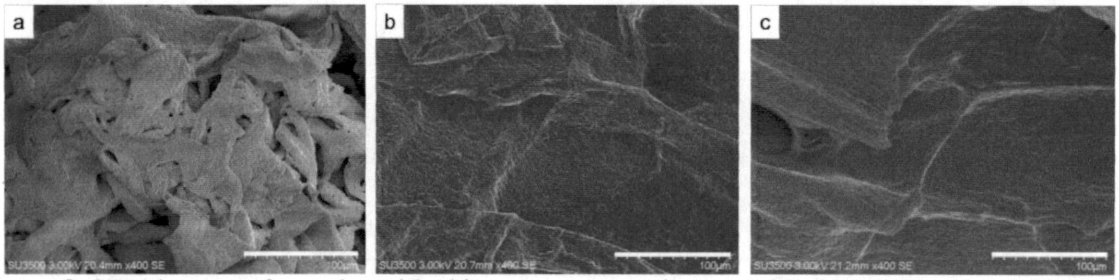

Figure 3. SEM images of samples with chitosan concentration: a) 1.0%, b) 1.5%, and c) 2.0% (scale bar represent 100 μm)

3.3 Swelling capacity

The swelling test was performed to examine the stability of freeze dried PECs in aqueous environment. After soaked in SBF for such intervals, it was found that PECs sample with chitosan concentration 1.0% could not maintain its shape and already swollen in 1 hour. On the other hand, PECs samples with chitosan concentration 1.5% and 2.0% could maintain their shape while absorbing water. After 2 hours, no more weight gaining were observed at those two samples. Thus, it could be said that the swelling maximum capacity of those samples reached in 2 hours. Swelling capacity of PECs sample with chitosan concentration 2.0% (3,332%) is higher than PECs sample with chitosan concentration 1.5% (2,823%). Therefore, it can be said that the higher the chitosan concentration in the mixture, the bigger the swelling capacity of scaffold could get.

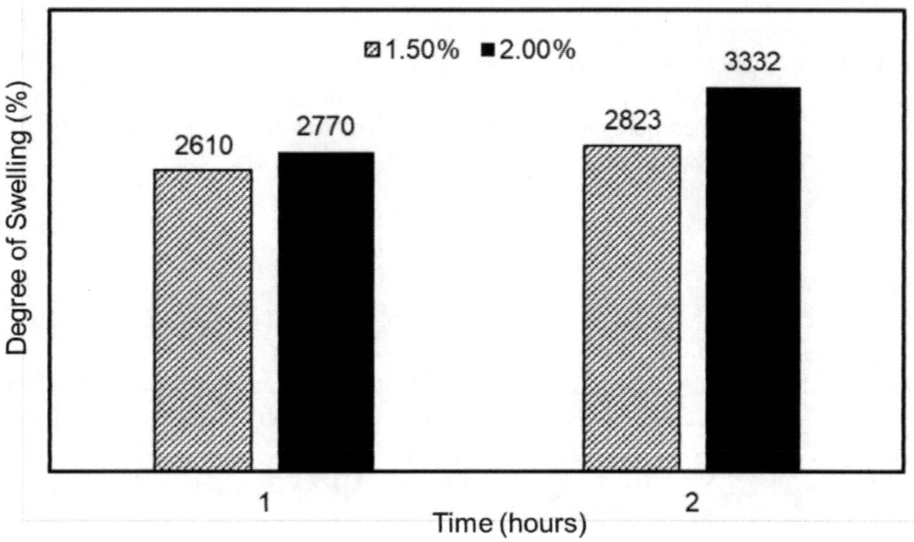

Figure 4. Swelling capacity of freeze drying sample with chitosan concentration 1.5% and 2.0% in aqueous environment

3.4Compressive Test

Mechanical properties of obtained samples were examined by compressive test and the results were presented in Fig. 5. As can be seen in Fig. 5, the ultimate compressive strength of the PECs samples increase as chitosan concentration increase. This trend happens due to the denser structure as the chitosan concentration increase (that were revealed by SEM observation), leading to the more strength when being compressed. Even though ultimate compressive strength of PECs sample with chitosan concentration 2.0% (1.22 kPa) still far from ultimate compressive strength of cancellous bone (1 – 12 MPa)[2], but PECs sample with chitosan concentration 2.0% showed higher ultimate compressive strength than sample of pure chitosan 2.0% (0.73 kPa), which means the PECs form is successfully enhance the mechanical strength of chitosan scaffold.

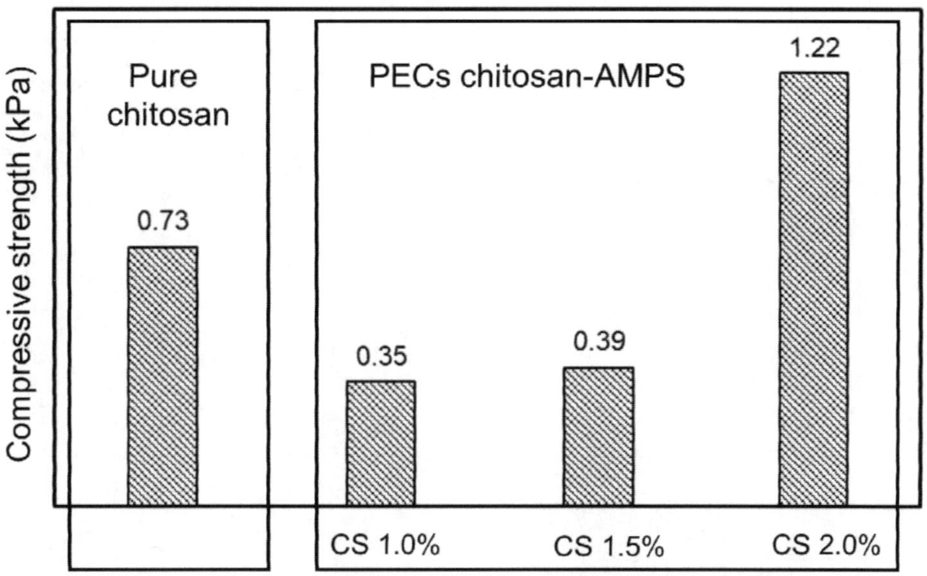

Figure 5. Ultimate compressive strength of pure chitosan 2.0% compared with PECs samples with chitosan concentration 1.0%, 1.5% and 2.0% respectively

4. Conclusion
Scaffold based on chitosan-PAMPS PECs has been successfully prepared using simple mixing method directly from their polymer solution. Based on swelling test and compressive test, PECs consist of chitosan 2.0% (w/v) and PAMPS 0.1% (w/v) shown the optimum stability in SBF medium amongst all PECs and highest ultimate compressive strength. Also, formation of PECs between chitosan and PAMPS is successfully enhance mechanical strength of chitosan scaffold, which is attractive initial information for future application of this material for bone implant scaffold.

5. References
[1] Amini, A. R., Laurencin, C. T., Nukavarapu, S. P. 2012 *Critical Reviews™ in Biomedical Engineering*, **40** 363 – 408
[2] Levengood, S. L., Zhang, M., 2014 Journal of Materials Chemistry B, **21** 3161–3184
[3] Li, Z., Ramay, H. R., Hauch, K. D., Xiao, D., Zhang, M., 2005 *Biomaterials* **26** 3919-3928.
[4] Venkatesan, J., Kim, S.-K. 2010 *Marine drugs* **8** 2252-2266
[5] Han, J., Zhou, Z., Yin, R., Yang, D., Nie, J., 2010 *International Journal of Biological Macromolecules* **46** 199-205
[6] Niknejad, H., Mahmoudzadeh, R. 2015 *Iranian Journal of Pharmaceuticals Research*, **14** 385 – 394
[7] Setoyama, H., Murakami, Y., Inoue, K., Iwata, H., Kitamura, H., Shimada, T., Kaji, H., Ikada, Y., Imamura, M. 1999 *Transplantation Proceedings* **31** 2818-2822
[8] Zhang, L., Wang, J., Ni, C., Zhang, Y., Shi, G. 2016 *Materials Science and Engineering C,* **58** 724 – 729
[9] Baxter, A., Dillon, M., Taylor, K. D. A., Roberts, G. A. F., 1992 *International Journal of Biological Macromolecules* **14** 166 – 169
[10] Kasaai, M. R. 2008 *Carbohydrate Polymers***71** 497 – 508
[11] Rodríguez-Vázquez, M., Vega-Ruiz, B., Ramos-Zuniga, R., Saldana-Koppel, D. A., Quinones-Olvera, L. F. 2015 *BioMed Research International* **2015** 821279
[12] Brugnerotto, J., Lizardi, J., Goycoolea, F. M., Arguelles-Monal, W., Desbrieres, J., Rinaudo, M 2001 *Polymer* **42** 3569 – 3580
[13] Sun, T., Xu, P. X., Liu, Q., Xue, J. A., Xie, W. 2003 *European Polymer Journal* **39** 189-192

Acknowledgements
This research was supported by International Collaboration Fund 2017 from Institut Teknologi Bandung.

Reduction the Oxygen Content of the Coconut Shell Char Produced by Using Simple Pyrolysis Method

M Rosi[1,*] and S Viridi[2]

[1]Instrumentation and Energy Engineering Group, School of Electrical Engineering, Telkom University, Jalan Telekomunikasi no 1 Terusan Buah Batu, Bandung
[2]Nuclear Physics and Biophysics Research Division, Faculty of Mathematics and Natural Sciences, Institut Teknologi Bandung, Jalan Ganesha no 10, Bandung

[*]Email: memoriarosi@telkomuniversity.ac.id

Abstract. Pyrolysis is known as a heating method which can decompose biomass to be the char in the absence of oxygen by providing carrier gas into the furnace. In practice, not all laboratories are equipped with carrier gas. Especially in large scale production, high cost production become the main problem. In this study, we propose simple pyrolysis method without carrier gas to produce coconut shell char as one of the biomass sources. The purpose of this research was to produce the coconut shell char using the simple pyrolysis method to obtain high purity carbon with low oxygen content. Experimental studies focused on the additional treatments complement to the pyrolysis process of the coconut shell. These treatments are proposed to minimize the oxygen content in the char. In this case, the oxygen bind to the carbon can reduce the carbon pore size. The produced coconut shell char were characterized by employing SEM-EDX, XRD, FTIR and Nitrogen Isotherm Physisorption. It was found that the treatments potential to reduce the oxygen content, contaminant and also increase the carbon yield.

1. Introduction

The use of biomass as renewable carbon source that is abundantly available in many country has been attractive for many industry such as purification, pharmaceutical, food industry, catalytic supports and energy devices [1-2]. In this case, the presence of low cost technology processing will be more advantageous, especially for large scale manufacturing. Here, we introduce a simple pyrolysis by means of conventional heating method using the muffle furnace with no carrier gas to decompose the biomass by burning off its chemical substance to be char for a longer time. The char can be converted to higher value products that can be used as activated carbon.

Coconut shell is one of the popular biomass with has high carbon yield and low ash [3-4]. Coconut shell has a higher volatile content that can produces carbon with large pore volume. Many literature report the experimental study of the coconut shell char by using pyrolysis with the carrier gas [5-8]. Unfortunately, according to our experience, the char produced with no carrier gas have low carbon yield and high oxygen content. To solve this problem, we propose the physicochemical treatments supplementary to the simple pyrolysis method. The physicochemical treatments consist of pre-treatment by microwave heating and acid (HCl) immersing and post treatment by acid (HCl re-immersing). As far as we know, we did not found similar treatment to reduce the oxygen content.

Meanwhile, highly oxygen content can reduce the char pore size and might degrade the char quality [9].

2. Experimental

2.1 Synthesis of coconut shell char

Coconut shell was taken from traditional market of Dayeuh Kolot, Bandung, Indonesia. The coconut shell were selected by the dark-shell color. The shell were cleaned by eliminate its fibrous and then crushed and sieve into equal size with range of 1-3 mm using hummer mill machine. After the preparation of coconut shell completed, all the sample were put into a muffle furnace (Carbolite, type 3216) and subsequently carbonized at 500°C with a heating rate of 10°C/min and held on for 2 hour. We varied four kind of the treatments as listed in Table 1.

Table 1. Physicochemical treatments to the coconut shell used in the simple pyrolysis.

Sample	Pre-treatment	Post-treatment
C-1	-	-
C-2	Microwave (1 kW) for 1 hour	-
C-3	Immerse in HCl 1 M for 12 hour, heating at 120°C for 1 hour and microwave (1 kW) for 1 hour	-
C-4	Immerse in HCl 1 M for 12 hour, heating at 120°C for 1 hour and microwave (1 kW) for 1 hour	Immerse in HCl 1 M for 12 hour and heating at 120°C for 1 hour

2.2 Characterization of coconut shell char

The resulted coconut shell char were characterized by employing Hitachi SU-3500 Scanning Electron Microscopy (SEM) to observe its surface morphology. The elemental composition of the char are analyzed with similar apparatus by using Energy Dispersive X-ray (EDX) spectroscopy method. FTIR spectroscopy (Thermoscientific Nicolete iS-10) was carried out to characterize the functional chemical groups of the char. X-ray diffractogram were obtained by Rigaku Smartlab (CuKα, $\lambda = 0.154$ nm) to investigate the char structure. Porous analyzing was used Nitrogen Isotherm Physisorption method (Quantachrome Autosorb Automated Gas Sorption). The surface area (SBET) was determined from its Nitrogen Isotherm Physisorption data using Brunauer-Emmett-Teller (BET) method. Total pore volume (VT) was defined as the volume of adsorbed N2 at a relative pressure of P/P0 = 0.99 [10]. The micropore volume (Vμ) was determined with the Dubinine-Radushkevich (DR) equation and the mesopore volume (Vm) was calculated by subtracting micropore volume from the total pore volume [11]. The pore size distribution of the char were obtained from Barrett, Joyner and Halenda (BJH method).

3. Results

Figure 1 represent SEM images of the produced coconut shell char. The chemical composition of the SEM images were determined using EDX as shown in Table 2. The char with no additional treatment (sample C-1) contains small bright particles among the porous carbon as given in Figure 1.(a). These particles is came from the impurity of K atom about 25.2 % which is already exists in the as-received coconut shell (Table 2). Figure 1.(b) demonstrated the larger and more dense carbon particle of sample C-2 compared to the sample C-1. The more dense carbon particle is consistent to the lower oxygen content. The porous structure seen in the sample C-3, as illustrated in in Figure 1.(c) is more opened up than that in sample C-4 (Figure 1.(d)). From the Table 1, it clear evidenced that the acid pre-treatment of sample C-2 and C-3 are successfully eliminate the K atom, while the acid post-treated of the sample C-4 can increase the carbon content and simultaneously reduce the oxygen content. The mechanism of the K atom elimination in the sample C-2 and C-3 might be analogous to the dissolution process of the mineral in the acid solution. The K atom which is easily trapped in the cellulose or lignin of the coconut shell would react with acids to form the potassium salts and water. The sample

C-4 which is treated by microwave might have a homogeneous heating so that the carbon loss and oxidation process is decrease along the process.

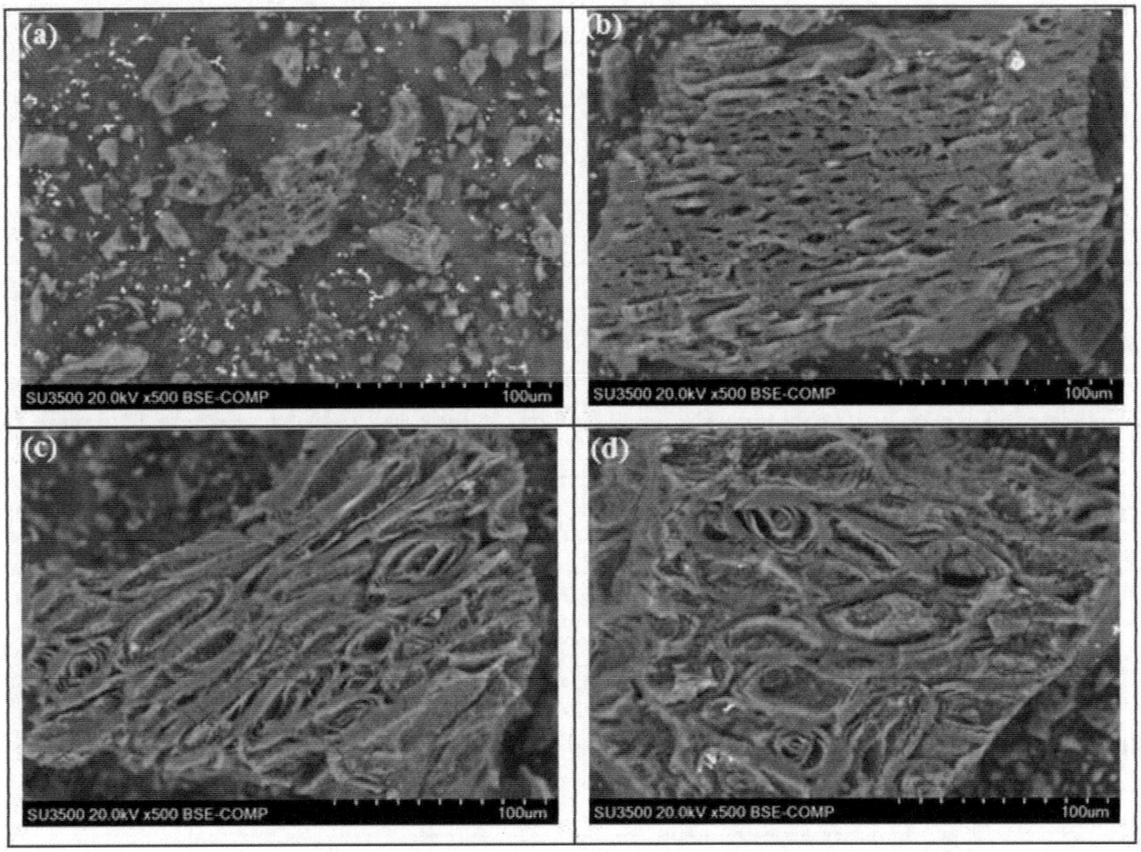

Figure 1. SEM images of coconut shell char:
(a) sample C-1, (b) sample C-2, (c) sample C-3, and (d) sample C-4.

Table 2. Atomic composition of the coconut shell char determined using EDX.

Sample	C (%)	O (%)	K (%)
C-1	49.3	25.5	25.2
C-2	52.0	15.0	33.0
C-3	78.0	22.0	0
C-4	81	19	0

IOP Conf. Series: Materials Science and Engineering **395** (2018) 012024 doi:10.1088/1757-899X/395/1/012024

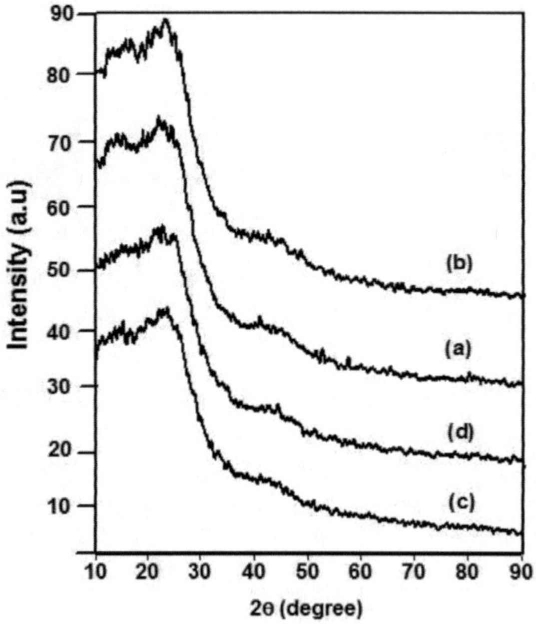

Figure 2. XRD analysis of coconut shell char:
(a) sample C-1, (b) sample C-2, (c) sample C-3, and (d) sample C-4.

X-ray diffraction of the char are displayed in Figure 2. The resulted char have an amorphous structure with widened peaks reflection between 11° and 30°. These results show that the lignocellulose structure of the char are destroyed [12]. The sample C-3 (Figure 2.(c)) has the widest reflection, indicates the more amorphous structure. Unfortunately, the XRD curve contain many noise, so that Refinement of the XRD patterns is not done here.

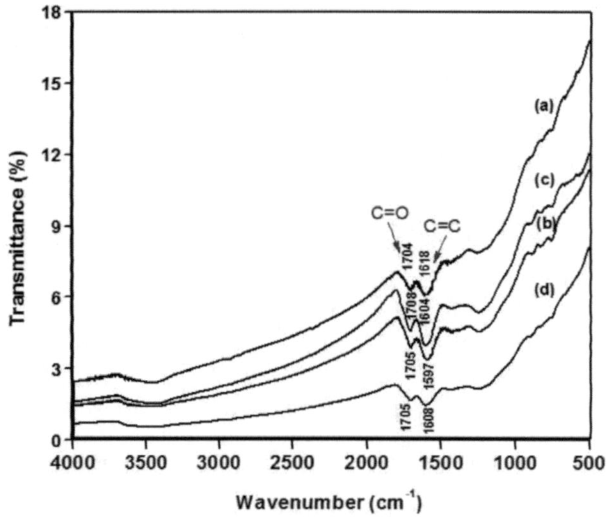

Figure 3. FTIR spectra of coconut shell char:
(a) sample C-1, (b) sample C-2, (c) sample C-3, and (d) sample C-4.

FTIR spectrum of the char is represented in Figure 3(a-d). Overall, there are two major peaks originated from C=O of carboxylic groups at 1708-1704 cm^{-1} and C=C bonds in aromatic at 1597-1618 cm^{-1} [8]. The C=O indicates the oxygen content in the char, whereas the C=C stretching attributed to the building block of carbon molecules of the char. The sample C-3 has the highest

IOP Conf. Series: Materials Science and Engineering **395** (2018) 012024 doi:10.1088/1757-899X/395/1/012024

intensity that show the highest density of carbon building block. This result is in line to that of EDX analysis (Table 2).

Table 3. Porous Characteristics of the coconut shell char.

Sample	S_{BET} (m^2/g)	V_T (cc/g)	V_μ (cc/g)	V_m (cc/g)	Average Pore Size (nm)
C-1	141	0.1	0.06	0.04	2.9
C-2	116.4	0.09	0.046	0.044	3.0
C-3	546.8	0.44	0.22	0.22	3.2
C-4	259.7	0.15	0.12	0.03	2.3

The porous characteristics of the char are given in Table 3. The highest specific surface area and highest pore volume is resulted from sample C-3 (S_{BET} = 546.8 m^2/g). This S_{BET} shows well pore development of the char which is comparable to the study reported in Ref 7. In contrary, the sample C-2 has lowest specific surface area of 116.4 m^2/g. This indicate that the acid pre-treatment can open up the char pore (micro- and meso pore). The higher S_{BET} of the sample C-1 is derived from its small particle size as represent in SEM images of Figure 1.(a). The higher specific surface area is consistent to the higher micopore volume. This suggest that the all of the char exhibited a highly microporous materials with slit-like pores [13]. By acid post-treatment, sample C-4 show the low specific surface area and small average pore-size compared to the sample C-3. It need further study to investigate this unexpected result.

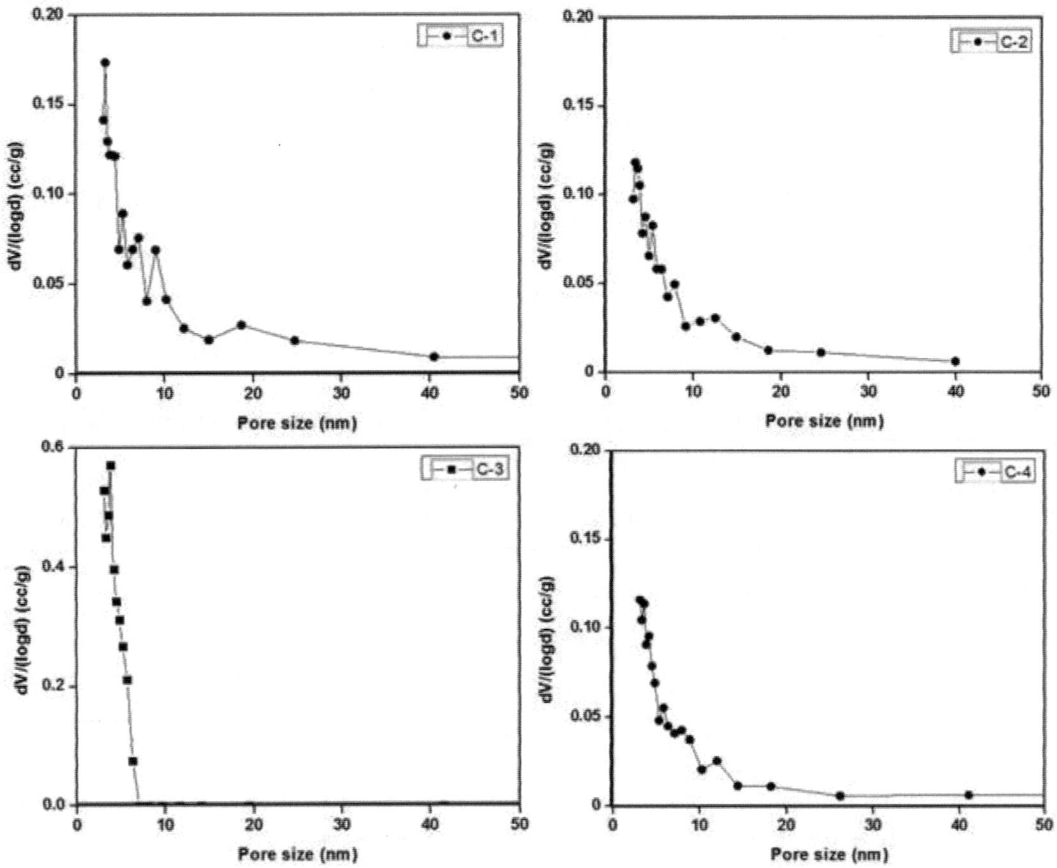

Figure 4. The BJH pore size distributions of the coconut shell char: (a) sample C-1, (b) sample C-2, (c) sample C-3, and (d) sample C-4.

The BJH pore size distributions of the char are demonstrated in Figure 4. The pores are distributed from micropores (with width less than 2 nm) and mesopores (with widths in the range of 2-50 nm width). All of the sample show the maximum pores is in the microporous range, except for sample C-3 has equal micro and mesopore volume. All of the sample have widely distributed pore in the range of 2-50 nm, except for sample C-3 which has limited pore-size in the range of 2-8 nm.

4. Conclusions

The coconut shell char have been manufactured by using simple pyrolysis method and comprehensively characterized. The SEM images shown the porous char structure as investigated from Nitrogen Isotherm Physisorption data and confirm by XRD with its amorphous structure. The acid and microwave heating treatments can increase the density of carbon in the char as shown by its SEM and EDX analyzing which is also supported from its FTIR spectra. Those results confirm that the higher carbon content is linear to the lower oxygen content. Therefore, the acid and microwave treatment can reduce the oxygen content of the coconut shell char.

5. References

[1] Yamada Y and Ozaki J-I 2004 Nanoporous Carbon *Encyclopedia of Nanoscience and Nanotechnology* vol 7 ed H S Nalwa (California: American Scientific Publishers) **pp** 237–262

[2] Yahya M A, Al-Qodah Z, and Ngah C W Z Agricultural bio-waste materials as potential sustainable precursors used for activated carbon production: A review 2015 *Renew. Sust. Energ. Rev.* **46** 218-235

[3] Su W, Zhou L, and Zhou Y Preparation of microporous activated carbon from coconut shells without activating agents 2003 *Carbon* **41** 861-863

[4] Das D, Samal D P, and Meikap Preparation of Activated Carbon from Green Coconut Shell and its Characterization 2015 *J. Chem. Eng. Process Technol.* 6 **5** 1-7

[5] Laine J, and Yunes S, Effect of the Preparation Method on the Pore Size Distribution of Activated Carbon from Coconut Shell 1992 *Carbon* (1992), 30 **4** 601-604

[6] Heschel W and Klose E On the sutability of agricultural by-products for the manufacture of granular activated carbon 1995 *Fuel* 12 **12** 1786-1791

[7] Achaw O-W, and Afrane G The evolution of the pore structure of coconut shells during the preparation of coconut shell-based activated carbons 2008 Microporous and Mesoporous Mater. 112 284-290

[8] Sundaram E G, and Natarajan E Pyrolysis of Coconut Shell: An Experimental Investigation 2009 *TJER* 6 **2** 33-39

[9] Tessmer C H, Vidic R D, and Uranowski L J Impact of Oxygen-Containing Surface Functional Groups on Activated Carbon Adsorption of Phenols 1997 *Environ. Sci. Technol.* 31 1872-1878

[10] Basta A H, Fierro V, El-Saied H, and Celzard A 2-Steps KOH activation of ricestraw: an efficient method for preparing high-performance activated carbons 2009 *Bioresour. Technol.* 100 3941-3947

[11] Stoeckli F, Daguerre E., and Guillot A, The development of micropore volumes ad widths during physical activation of various precursors 1999 *Carbon* 37 2075-2077

[12] Zubrik A, Matik M, Hredzák S, Lovás M, Danková Z, Kováčová M, and Briančin J Preparation of chemically activated carbon from waste biomass by single-stage and two-stage pyrolysis 2016 *J Clean Prod.* (manuscript)

[13] Sing K S W, Everett D H, Haul R A W, Moscou L, Pierotti R A, Rouquerol J, and Siemieniewska T Reporting Physisorption Data for Gas/Solid Systems 1985 *Pure Appl. Chem.* 57 **4** 603-619

IOP Conf. Series: Materials Science and Engineering **395** (2018) 012024 doi:10.1088/1757-899X/395/1/012024

Acknowledgments
The authors wish to thank to The Ministry of Technology Research and Higher Education (Kemenristek-Dikti), Republic of Indonesia for its financial support of our research through a pasca doctoral grant (Hibah Pasca Doktor) for 2017-2018. The first author also thank to the Institut Teknologi Bandung for research collaboration.

IOP Conf. Series: Materials Science and Engineering **395** (2018) 012025 doi:10.1088/1757-899X/395/1/012025

The Effect of Paramagnetic Agent on Relaxation and Mechanical Properties of Agar Hydrogel for Phantom in Magnetic Resonance Imaging

Y Dwihapsari*, N Maulidiyah, Darminto

Department of Physics, Institut Teknologi Sepuluh Nopember Surabaya (ITS), Kampus ITS Sukolilo Surabaya 60111, Indonesia

*yanuritadh@physics.its.ac.id, yanuritadh@gmail.com

Abstract. Agar hydrogel is largely used in magnetic resonance imaging study as tissue mimicking phantom material for its cost-effectiveness and easy preparation. Spin-spin relaxation time, T_2 is one important parameter in Magnetic Resonance Imaging study. However, T_2 of agar sample is higher than those of human tissues. The addition of paramagnetic agent such as copper sulphate or $CuSO_4$ had been observed to match T_2-value of human tissues. In this study, the effect of $CuSO_4$ on relaxation and mechanical properties of agar is observed.

Keywords: hydrogel, agar, magnetic resonance imaging, mechanical properties, shear modulus, T_2, relaxation time, $CuSO_4$, phantom

1. Introduction

Agar hydrogel is largely used in magnetic resonance imaging study for tissue mimicking phantom material for its cost-effectiveness and easy preparation. In order to make a material eligible for tissue mimicking phantom material in Magnetic Resonance Imaging (MRI) study, the properties of material must serve some important parameters in MRI, such as spin-lattice relaxation time (T_1) and spin-spin relaxation time (T_2). T_2 refers to relaxation time of water protons which describes cumulative loss in phase coherence and results in signal decay from each voxel during MRI acquisition. The T_2-relaxation results from transient magnetic field due to molecular motion and depends on chemical exchange inside the material.

The increasing agar concentration results in lower T_2 although it is relatively high compared to T_2 of some human tissues [1]. The paramagnetic agent is normally added to get much lower T_2 and suitable with T_2 of corresponding tissues. The previous study showed that addition of paramagnetic agents, such as $CuSO_4$ and $Mn(NO_3)_2$ reduce T_2 along with increasing concentration of agar or other hydrogels [1, 2]. Several agar hydrogel studies have observed the mechanical properties of agar and showed that increasing agar concentration could increase the mechanical properties of agar hydrogel [3].

In this study the effect of concentration of agar and $CuSO_4$ on spin-spin relaxation time (T_2) and its shear modulus are discussed. The experiments of agar with various concentrations were measured ca. 2 h after gelation and scanned using MRI to obtain T_2. The shear modulus was observed using dynamic mechanical analyser (DMA) tool. The relation between shear modulus and T_2 is observed and plotted to explore their link to concentration of agar and $CuSO_4$.

IOP Conf. Series: Materials Science and Engineering **395** (2018) 012025 doi:10.1088/1757-899X/395/1/012025

2. Materials and methods

The solution of agar hydrogel was formed by mixing agar powder and 50 ml distilled water and boiled for 20 min to form 2.5, 5, and 7.5 w/v% of agar gel. The paramagnetic agent $CuSO_4$ was added to agar and distilled water to form concentration of 0.4 and 5.9 w/v% and boiled to form hydrogel. The solution was placed into plastic tube and cooled down to form hydrogel in room temperature. The samples were positioned in such a row and scanned after 2 – 4 h of preparation for measurement of spin-spin relaxation time T_2.

The MRI experiment to measure T_2 was performed using Carr-Purcell Meiboom Gill (CPMG) sequence in 1.5 T MRI scanner (Signa, General Electric, US) with spine gradient coil. Five slices were taken from each samples. The experiment was performed with parameters: field-of view (FOV) = 100 mm × 100 mm, matrix size = 64 × 64, repetition time (TR) = 1000 – 2000 ms, echo-time (TE) = 10 – 200 ms, and slice thickness = 5 mm. The image processing and analysis software, ImageJ was used to obtain signal intensity. T_2-value was obtained by fitting mean signal intensity into TE.

Each sample was cut into 5 mm x 5 mm x 1 mm plates for measurement of mechanical properties. The shear modulus of agar hydrogel was observed using DMA Mettler Toledo SDTA861 by varying force load from 1 to 20 N. The shear modulus was acquired by plotting and analysing stress-strain relation curve. The shear modulus of each sample was plotted against T_2 to obtain the relation between relaxation time and mechanical properties of agar hydrogel.

3. Results and discussion

Spin-spin relation time T_2 from various concentration of agar and mixture of agar and $CuSO_4$ is shown in Table 1. The result shows the decreasing T_2 as agar concentration increases, even less with addition of $CuSO_4$. The value of T_2 in our experiment is comparable to T_2-value from other experiments [1] and adjustable to T_2 of human tissue from other studies which range between 30 – 100 ms at 1.5 T [4 - 6]. The small reduction of T_2 due to increasing agar concentration shows the effectiveness of increasing $CuSO_4$ concentration for reducing T_2 value. As shown in Table 1, the addition of $CuSO_4$ affects T_2-value for the same agar concentration. The reduction of T_2 as a result of increasing $CuSO_4$ concentration is up to 50% for 0.4% and only 40% for 5.9w/v% of $CuSO_4$. It shows that the addition of $CuSO_4$ yields the maximum reduction in particular range of concentration. The results in this study are relevant with results from other study which found linear reduction in T_2 when agar and $CuSO_4$ concentration are increased [1, 2]. The linear reduction of T_2 found in the study is useful in predicting the concentration of paramagnetic agents needed for preparation of tissue-mimicking phantom materials.

Table 1. Spin-spin relaxation time, T_2 of various concentration of agar and mixture of agar and $CuSO_4$.

Sample	$CuSO_4$ concentration (w/v%)	Agar concentration (w/v%)	T_2 (ms)	SD
1	0	2.5	116.705	4.119
2	0.4	2.5	51.493	2.770
3	5.9	2.5	32.798	4.630
4	0	5	102.271	18.818
5	0.4	5	50.025	2.538
6	5.9	5	24.149	2.899
7	0	7.5	90.799	44.139
8	0.4	7.5	46.211	2.415
9	5.9	7.5	23.998	2.604

The addition of $CuSO_4$ concentration give insignificant contribution to shear modulus in the same agar concentration as described in Table 2. In the experiment of shear modulus, data from 2.5% agar concentration was excluded due to broken sample during testing. The significant increase was found in higher agar concentration and showed that shear modulus is significantly dependent on agar concentration.

Table 2. Shear modulus of various concentration of agar and mixture of agar and $CuSO_4$. Agar 2.5w/v% was excluded due to broken sample during testing.

Sample	$CuSO_4$ concentration (w/v%)	Agar concentration (w/v%)	Shear modulus (kPa)
1	0	5	178.225
2	0.4	5	150.180
3	5.9	5	131.475
4	0	7.5	380.449
5	0.4	7.5	365.749
6	5.9	7.5	302.882

The link between T_2 and shear modulus of agar hydrogel is shown in Figure 1. The addition of $CuSO_4$ significantly reduces T_2 but insignificantly affects shear modulus. On the contrary, increasing agar concentration significantly increases shear modulus but gives little effect to T_2-value. The linear relation is found between T_2 and shear modulus and the fitting shows that the slope depends on agar concentration and can be determined with more variation of agar concentration.

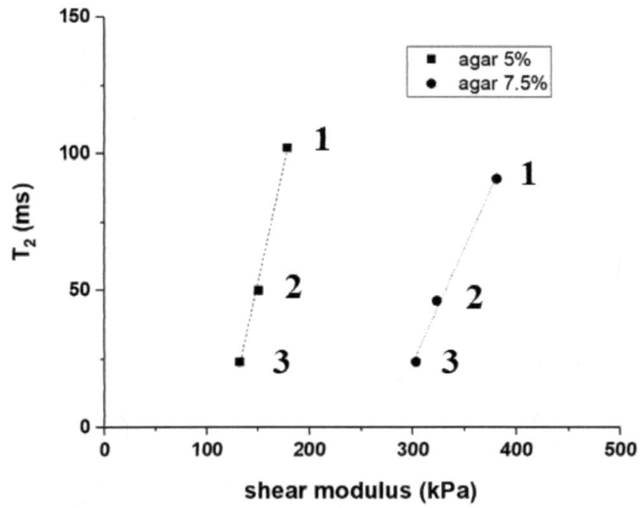

Figure 1. Shear modulus vs T_2 of agar and mixture of agar and $CuSO_4$. 1) Agar + $CuSO_4$ 0%, 2) Agar + $CuSO_4$ 0.4%, 3) Agar + $CuSO_4$ 5.9%. The result from agar 2.5% was excluded in this plot.

4. Conclusion

The effect of adding paramagnetic agent, in our study $CuSO_4$ to agar hydrogel had been observed. The results show that addition of $CuSO_4$ results in significant reduction in T_2-value but insignificant increase in shear modulus of agar hydrogel. The shear modulus is linearly related to T_2-value along with increasing concentration of agar and $CuSO_4$. The study works for limited concentration of agar and more variation of agar and $CuSO_4$ concentration is needed to observe the link between T_2 and shear modulus in agar hydrogel.

5. References

[1] Mitchell MD, Kundel HL, Axel L, Joseph PM 1986 *Magn Reson Imaging* 4(3) 263-266
[2] Hellerbach A, Schuster V, Jansen A, Sommer J 2013 *Plos One* **8**(8) 1 - 8
[3] Ahearne M, Yang Y, El Haj AJ, Then KY, & Liu K-K 2005 *J. of the Royal Society Interface* **2**(5) 455–463
[4] Sled JG, Pike GB 2001 *Magn Reson Med* **46** 923–932
[5] Gold GE, Han E, Stainsby J, Wright GA, Brittain J, Beaulieu C 2004 *Am J Neuroradiol* **183** 343–350
[6] Pell GS, Briellmann RS, Waites AB, Abbott DF, Jackson GD 2004 *Neuroimage* **21** 707–713

Acknowledgments

ITS is highly acknowledged for funding Research Grant Penelitian Pemula ITS 2017. Authors wish to thank Prof Suminar Pratapa and Nur Aini Fauziyah for discussion and assistance on DMA measurement.

IOP Conf. Series: Materials Science and Engineering **395** (2018) 012026 doi:10.1088/1757-899X/395/1/012026

Possible Weak Ferromagnetism in $Pr_{2-x}Ce_xCuO_4$ Nanocrystals at Normal State

B Triono, P E D Putra, R Irfanita, M A Baqiya*, Darminto

Department of Physics, Institut Teknologi Sepuluh Nopember (ITS), ITS Campus, Keputih-Sukolilo, Surabaya 60111, Indonesia

* Corresponding Email: malikabits@physics.its.ac.id

Abstract. Nano-crystals of $Pr_{2-x}Ce_xCuO_4$ (PCCO) with x = 0 and 0.15 have been successfully synthesized by chemically dissolved method employing HNO_3 as dissolving agent. The PCCO samples were calcinated at 900 – 1000°C for 5 – 30 h. The PCCO samples were characterized by x-ray diffraction (XRD) and vibrating sample magnetometer (VSM) for investigating the phase formation and crystals size, and for studying the magnetic properties at normal state, respectively. Based on the XRD data, PCCO with x = 0.10 and 0.15 calcinated at temperature of 1000°C for 15 h show no impurity phases with crystal size of around 50 nm using Rietveld analysis. It is shown that the non-superconductive PCCO nano-crystals have very weak ferromagnetic behavior at room temperature. This phenomenon is probably due to magnetic moments rising from oxygen vacancies of the nanocrystals surface.

1. Introduction

Researches on synthesis and magnetic properties of cuprate superconductors have shown a rapid development in both hole- and electron-doped cuprates. Generally, a micrometer particles and even single crystal bulk superconducting samples have been intensively produced in order to investigate the basic physical properties of the materials [1]. However, it is very rare to find superconducting materials in nanometer crystal size. There are some common methods to synthesis superconducting materials, such as solid state reaction [2-4] and ball milling [5-7], in which both synthesis techniques produce a sample with micrometer particle size (> 100 nm). In the case for producing a nanometer crystal size, precipitation method or dissolved method using NaOH, HCl and HNO_3 as dissolved agents have been successfully conducted [8-11].

Cuprate based superconductors with nanometer crystal size have been shown to have unique property so-called room-temperature ferromagnetism (RTFM). Shipra *et al.* [12] have shown that $YBa_2Cu_3O_7$ (YBCO) with crystal size below 100 nm shows ferromagnetic properties at above critical temperature, T_c, as well as at room temperature. This phenomenon is indicated by the appearance of magnetic hysteresis at the normal state. However, in the micrometer bulk samples, it has been shown that a paramagnetic behavior was observed clearly [13]. The presence of ferromagnetism is expected due to the oxygen vacancies (crystal defects) on surface of the nanoparticles producing weak and localized magnetic moments. The RTFM phenomenon has also been reported by M. A. Baqiya and co-workers in the $Bi_2Sr_2CaCu_2O_8$ (Bi-2212) nanoparticles. They have observed the existence of RTFM in the Bi-2212 nanoparticles prepared by dissolved method with crystal size less than 200 nm [8].

This paper is aimed to study the magnetic properties at normal state of T'-$Pr_{2-x}Ce_xCuO_4$ (T'-PCCO) nanocrystals prepared by chemical dissolved method. As it is known that the physical properties of the T'-structure are strongly influenced by the presence of apical oxygen [14]. A small number of oxygen vacancies either on the apical site or the other oxygen sites may affect the overall magnetic properties of the cuprates. Therefore, it is also important to investigate the magnetic behaviour of the PCCO nanocrystals at normal state.

2. Experimental

$Pr_{2-x}Ce_xCuO_4$ (PCCO) nanocrystals with $x = 0$ and 0.15 were prepared by dissolved method with HNO_3 as the dissolving agent. The dissolved method used in this present research was similar with previous works [8]. Pr_6O_{11}, CeO_2, and CuO powders with high purity (99.9%) have been used as the raw materials. First, the raw materials were dried and weighted according to the desired stoichiometry. The dried raw materials were dissolved by HNO_3 separately at around 150°C and then mixed them together until it forms a precipitate. The step was continued by performing pre-calcination process for the precipitate at 400°C for 4 h. Finally, the precursor powder was calcinated at 900 – 1000°C for 5 – 30 h. The structural properties and magnetic hysteresis curves of the PCCO samples were measured by XRD and VSM, respectively. Phase identification and crystal structure of the samples were investigated using XRD with scanning angle 2θ ranging from 10° to 70° and operated at 40 kV and 30 mA with CuKα_1 radiation with wavelength of 1.54 Å. The lattice parameter and the average crystal size were estimated by Rietveld analysis using Rietica [15] and materials analysis using diffraction (MAUD) [16] software, respectively. VSM (Dexing Magnet Ltd., VSM 250) measurements were conducted with external magnetic field ranging from -2 T to 2 T at room temperature (23°C) resulting magnetic hysteresis curves.

3. Results and Discussion

Figure 1(a) and 1(b) show the XRD patterns of PCCO nanocrystals with $x = 0$ and 0.15 at various calcination temperatures, respectively. Based on the phase identification, all peaks are detected as PCCO phase with PDF#080-2368. However, at the calcination temperature of 900°C, there is an impurity phase of Pr_6O_{11}. Matsuda *et al.* [17] reported that the single phase of 214 systems can be obtained using solid state reaction technique at 1030°C. In this research, the PCCO phase has been formed at calcination temperature of 1000°C for 15 h without any impurities. In fact, the formation of PCCO phase has been initiated at 900°C with small amount of impurities less than 15%. This means that the dissolved method used in this present research has successfully formed the 214 phase (PCCO phase) at a lower temperature than 1030°C. This is due to the dissolved method is one of bottom up synthesis method which give the particles easily to react each other even at relatively low temperature.

Figure 1(c) shows peak shift between PCCO nanocrystals without and with Ce doping. All diffraction peaks shift to larger diffraction angle with the addition of Ce doping meaning that all Ce atoms were successfully doped into the crystal. The values of lattice parameter and crystal size of PCCO samples obtained from the analysis results of Rietica and MAUD software, respectively, are shown in Table 1. PCCO phase (ICSD#72246 with tetragonal structure and space group I4/*mmm*) was used for Rietveld analysis achieving the goodness of refinement (GoF) less than 2%. It can be seen that the addition of Ce doping decreases the c-axis length. This is consistent with the former research by Matsumoto *et al.* [18] that there is doping dependence of lattice parameter in the cuprate system. The ionic radius of Ce^{4+} (1.11 Å) is smaller than that of Pr^{3+} (1.27 Å). Therefore, the c-axis length shrinks in the doped sample because the Pr^{3+} ions have been replaced by Ce^{4+} ions perfectly [18].

Based on analysis results in Table 1, it is found that PCCO samples with $x = 0$ and 0.15 have crystal size of 40 – 118 nm. These results show that the calcination temperature and the holding time affect the crystal size due to the thermal energy for resulting larger crystal size. As for the effect of doping, Table 1 shows that a- and b-axis lengths do not change significantly with additional Ce doping and different calcination time. However, c-axis length increases with the increase of calcination time because the longer calcination time provides more thermal energy to expand the lattice parameter (c-axis length).

IOP Conf. Series: Materials Science and Engineering **395** (2018) 012026 doi:10.1088/1757-899X/395/1/012026

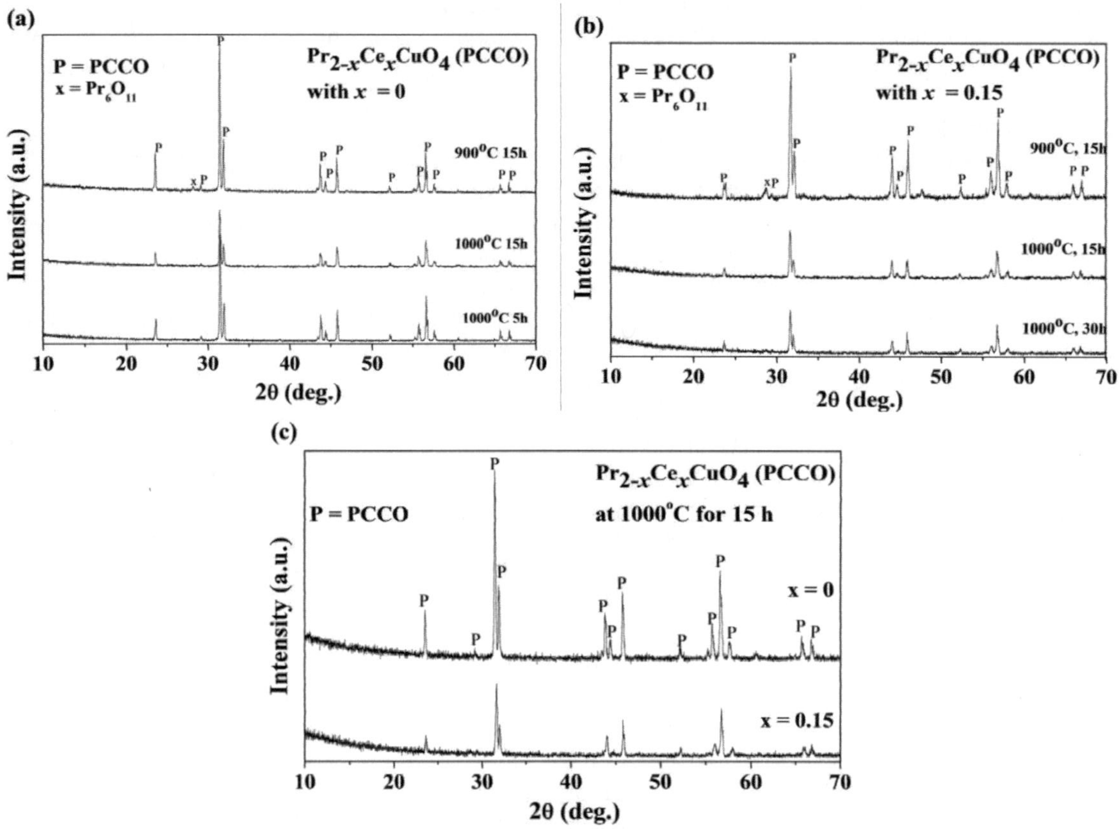

Figure 1. XRD patterns of PCCO nanocrystals with different calcination process and holding time: 900 - 1000°C for 5 - 30 h. (a) $x = 0$ and (b) $x = 0.15$; (c) Comparison between the XRD patterns of single phase PCCO with $x = 0$ and 0.15 (calcination at 1000°C for 15 h).

Table 1. The results of crystal analysis of PCCO samples with $x = 0$ and 0.15 using Rietveld analysis.

PCCO Samples	Calcination Temperature	Holding Time (h)	Crystal size (nm)	Lattice Parameter (Å)	
				a=b axis	c axis
$x = 0$	1000°C	5	118 ± 20	3.903(0)	12.058(1)
	1000°C	15	61 ± 1	3.904(0)	12.062(1)
$x = 0.15$	1000°C	15	40 ± 1	3.908(1)	12.012(2)
	1000°C	30	46 ± 1	3.902(0)	12.025(1)

Figure 2 shows magnetic hysteresis curves for PCCO nanocrystals with $x = 0$ and 0.15 measured at room temperature. Both samples are non-superconducting samples and show weak RTFM phenomena. This result is in accord to the former paper by Shipra *et al.* [12]. It can be seen that the hysteresis curves show no magnetic saturation up to 1 T representing dominant paramagnetic properties in both PCCO samples. Based on Figure 2, PCCO nanocrystals without Ce doping shows a higher value of paramagnetic susceptibility compared to the doped one. It seems that the increase of Ce doping results in the decrease of magnetic hysteresis and turns into relatively weak ferromagnetic at room temperature. This is probably due to the contribution of Ce^{4+} magnetic moments affecting the overall magnetization in the doped PCCO.

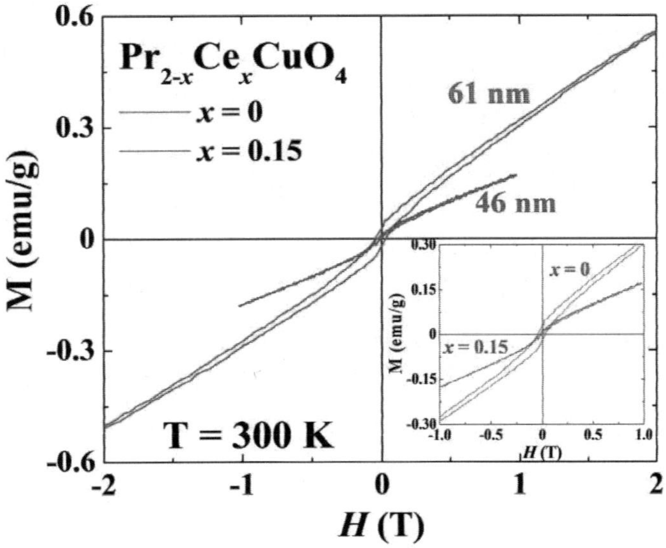

Figure 2. Magnetic hysteresis curves for PCCO nanocrystals with $x = 0$ and 0.15 at room temperature.

Both undoped and doped PCCO samples with crystal size less than 100 nm exhibit RTFM feature. The crystal size dependence on the magnetic properties of cuprates has been reported by S. K. Hasanain *et al.* [19]. The higher crystal size of the nanoparticles results in a smaller magnetization and finally becomes paramagnetic. Moreover, the increase of Ce doping in the PCCO crystal may be responsible for the disturbance of net magnetic moments leading to a competition between magnetic moments rising from Ce ions and oxygen vacancies. However, the appearance of RTFM in PCCO nanocrystals should be investigated in more detail to elucidate the correlation between crystal size effect and magnetic moments rising from oxygen defects.

4. Summary

PCCO nanocrystals with $x = 0$ and 0.15 have been successfully synthesized by chemically dissolved method. The XRD patterns show single phase of PCCO crystal obtained by calcination at 1000°C for more than 15 h. Based on Rietveld analysis, it is obtained the smallest crystal size of about 40 nm. The lattice parameters (a- and b-axis length) do not change significantly with different calcination temperatures. However, the c-axis length decreases with addition of Ce doping. Both PCCO nanocrystals with $x = 0$ and 0.15 have shown very weak RTFM behavior. It is found that PCCO nanocrystal with $x = 0$ shows a relatively strong magnetization without any magnetic saturation compared with $x = 0.15$. These results suggest that both PCCO nanocrystals have dominant paramagnetic properties with very weak RTFM feature.

Acknowledgement

Thanks for all members of superconductor team at Koike Laboratory, Tohoku University, Japan who help experimental works and other activities related to this project. This research was partly support by Indonesia Endowment Fund for Education (LPDP) scholarship (B Triono) and grand project "Penelitian Unggulan Perguruan Tinggi ITS Surabaya" from Ministry of Research, Technology and Higher Education of the Republic of Indonesia (M A B).

References

[1] Rao, C.N.R., Nagarajan, R., Vijayaraghaven, R. 1993 *Supercond. Sci. Tech.* **6**(1) 1

[2] Pignon, B., Veron, E., Noudem, J., Ruyter, A., Ammor, L., Monot-Laffez, I. 2006 *Physica C* **434**(1) 45-52
[3] Sözeri, H., Özkan, H., Ghazanfari, N. 2007 *J. Alloy. Compd.* **428**(1) 1-7
[4] Warner, T.E. 2011 *In: Synthesis, Properties and Mineralogy of Important Inorganic Materials*. pp. 144-164. John Wiley & Sons, Ltd
[5] Kruaehong, T., Sujinnapram, S., Nilkamjon, T., Ratreng, S., Udomsamuthirun, P. 2017 *J. Aust. Ceram. Soc.* **53**(1) 3-10
[6] Yavuz, M., Maeda, H., Vance, L., Liu, H.K., Dou, S.X. 1998 *Supercond. Sci. Tech.* **11**(10) 1153
[7] Hamrita, A., Ben Azzouz, F., Madani, A., Ben Salem, M. 2012 *Physica C* **472**(1) 34-38
[8] Baqiya, M.A., Widodo, H., Rochmawati, L., Darminto, Adachi, T., Koike, Y. 2012 *AIP Conf. Proc.* **1454**(1) 260-263
[9] Astuti, F., Baqiya, M.A., Darminto 2013 *AIP Conf. Proc.* **1554**(1) 97-100
[10] Wang, A.Q., Han, H.H., Li, M., Xie, J.P., Wu, Q.J., Ma, D.Q. 2015 *Key Eng. Mater* **645-646**, 145-150
[11] Jian-feng, H., Xie-rong, Z., Li-yun, C., Xin-bo, X. 2009 *J. Mater. Process. Tech.* **209**(6) 2963-2966
[12] Shipra, Gomathi, A., Sundaresan, A., Rao, C.N.R. 2007 *Solid State Commun.* **142**(12) 685-688
[13] Sundaresan, A., Rao, C.N.R. 2009 *Nano Today* **4**(1), 96-106
[14] Fournier, P. 2015 *Physica C* **514** 314-338
[15] Rietveld, H. 1969 *J. Appl. Crystallogr.* **2**(2) 65-71
[16] Lutterotti, L., Scardi, P. 1990 *J. Appl. Crystallogr.* **23**(4) 246-252
[17] Matsuda, M., Endoh, Y., Hidaka, Y. 1991 *Physica C* **179**(4) 347-352
[18] Matsumoto, O., Utsuki, A., Tsukada, A., Yamamoto, H., Manabe, T., Naito, M 2009 *Physica C* **469**(15) 924-927
[19] Hasanain, S.K., Akhtar, N., Mumtaz, A 2011 *J. Nanopart. Res.* **13**(5) 1953-1960

Institute of Physics
Dirac House, Temple Back
Bristol BS1 6BE UK

ISSN: 1757-8981
ISBN 978-1-5108-7025-3